Light Emitting Diodes

by

Forrest M. Mims III

HOWARD W. SAMS & CO., INC.
THE BOBBS-MERRILL CO., INC.
INDIANAPOLIS • KANSAS CITY • NEW YORK

FIRST EDITION

SECOND PRINTING—1976

International Standard Book Number: 0-672-20964-0
Library of Congress Catalog Card Number: 73-75078

Contents

CHAPTER 4

CHAPTER 5

CHAPTER 6

Preface

In the spring of 1965 I visited Mr. Edward Bonin of Texas Instruments in Dallas in hopes of obtaining some sample infrared-emitting diodes. At that time, commercial light emitting diodes (LEDs) were available, but they were very expensive. I had never seen an LED until Mr. Bonin pulled a small device from a desk drawer and pressed a switch. A bright red glow emerged from a gold-plated metal can about the size of a transistor, and ever since I've been fascinated by light-emitting semiconductors.

LEDs are experiencing important new developments, and in recent years the field has literally exploded. Business is booming, largely as a result of the demand for LED displays in electronic calculators and digital equipment, and millions of LEDs are manufactured each month. New developments, particularly in the field of visible emitters, occur frequently.

This book could not have been written without the invaluable assistance of many individuals and companies. John W. Wall II, of General Electric Company Miniature Lamp Department, provided numerous sample LEDs, photographs, and encouragement. Dick Lundgren, also of GE, secured permission to reprint information contained in the company's *Solid State Lamp Manual*. Dr. Richard Glicksman of RCA supplied technical information, advice, and samples of calibrated LEDs. Dr. G. W. Paxton, president of Spectronics, Inc., provided technical advice and a photographic tour of his plant.

Phil Davis, vice-president of United Detector Technology, provided encouragement, phtographs, and advice about LED detectors. H. Edward Roberts, president of Micro Instrumentation and Telemetry Systems, Inc., permitted me to photograph LED displays used by his company in the manufacture of digital calculators and clocks. Mr. Roberts also granted permission to reproduce the circuit diagrams for the Opticom LED communicator described in Chapter 8.

Monsanto sales representative Dwayne Frye provided invaluable assistance in obtaining samples of new LEDs and literature. Jack Pintar provided information and samples of Litronix LEDs. Hewlett-Packard's Jim Graham supplied an assortment of his firm's LEDs and technical literature. Ken Horton, Bob Cole, Lin Wetterau, and Ted Meador of Texas Instruments provided technical advice, literature, photographs, and sample LEDs.

Infrared-sensitive phosphor screens were supplied by C. H. Fairbanks of Kodak and by Optical Engineering Company. F. S. Keith of RCA supplied an infrared image-converter tube. Walter Lekki of Optical Coating Laboratory provided infrared interference filters and literature. James P. Miller of the Mathematics Department of the University of New Mexico and president of the Southwest Research Association provided valuable assistance with the optical radar range equations in Chapter 8.

Finally, I am most appreciative of the weeks of labor contributed by my wife Minnie in proofreading and typing the drafts and final manuscript. Her understanding, patience, and help greatly simplified the project. To these many individuals, companies, and my loving wife I offer my sincere appreciation and thanks.

FORREST M. MIMS, III

CHAPTER 1

Semiconductor Light Emission

In 1907 electrical engineer Henry J. Round touched two wires connected to a battery to a crystal of silicon carbide found near Niagara Falls. As he subsequently reported: "On applying a potential of ten volts between two points on a crystal of carborundum, the crystal gave out a yellowish light." Though he may not have realized it at the time, Round had operated the first crude solid-state lamp, the light emitting diode.

Round's historic experiment can be easily duplicated with the help of some silicon carbide crystal fibers. The crystals are held fast in an alligator clip connected to from 10 to 30 volts dc and touched with a wire electrode that is also connected to the power supply. Very bright flashes of yellow light are produced. Fig. 1-1 is a photographic time exposure showing more than a dozen such flashes.

The discovery of light emission in silicon carbide was important for several reasons, but the most significant reason concerned the mechanism that caused the carborundum crystal to emit light. Most electric light sources produce visible radiation by means of incandescence of a hot electrode, luminescence of a glowing plasma, or fluorescence of a phosphor coating. Depending on the filament temperature, incandescent sources such as the tungsten filament lamp emit visible light with an efficiency of about 5 percent. The total radiated output of the lamp extends from the visible region well into the infrared, as shown in Fig. 1-2, and varies with the filament temperature.

Luminescent and fluorescent sources generally emit a broad visible spectrum punctuated by one or more narrow-bandwidth, high-intensity peaks or spikes. The spikes are characteristic of the atomic structure of the gas or phosphor light source, and Fig. 1-3 shows the spectrum from a xenon arc lamp with several such peaks.

Most of the electric light sources that were in use in Round's time are still practical. Each is well suited for general illumination, and some lamps are capable of producing exceedingly high luminous outputs.

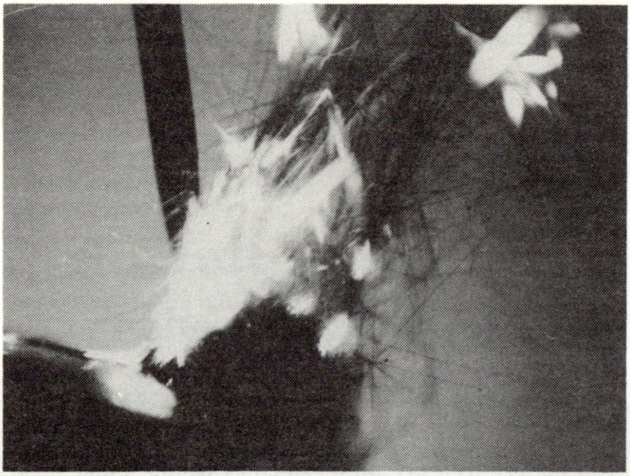

Fig. 1-1. Electroluminescence in silicon carbide crystal fibers (see text).

But each of the conventional light sources is unsuited to a broad range of potential applications because of their slow response time, inherent fragility, and limited lifetime.

The light emitting diode (LED) conveniently eliminates the limitations of conventional lamps with its solid-state reliability, speed, and compact size. While semiconductor light sources are not yet suited for general-purpose illumination, their operating characteristics make them well suited for a host of practical applications, including miniature panel indicators, digital readouts, light-beam communications, optical radar, and logic status indicators. Since they emit a narrow-wavelength spectrum, various colored sources can be fabricated without the need for filters. The output spectra of several LEDs is shown superimposed

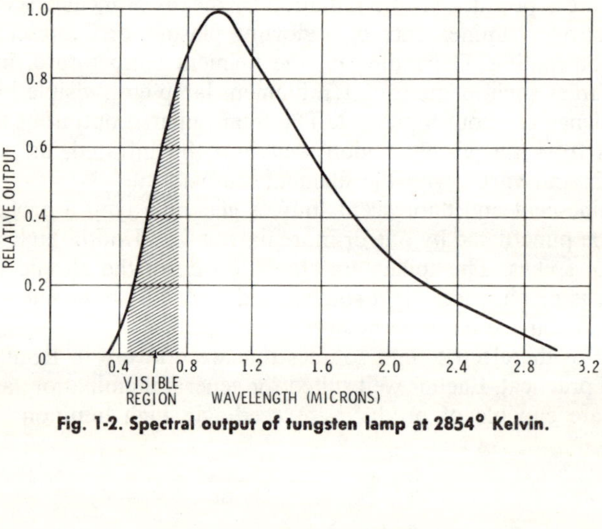

Fig. 1-2. Spectral output of tungsten lamp at 2854° Kelvin.

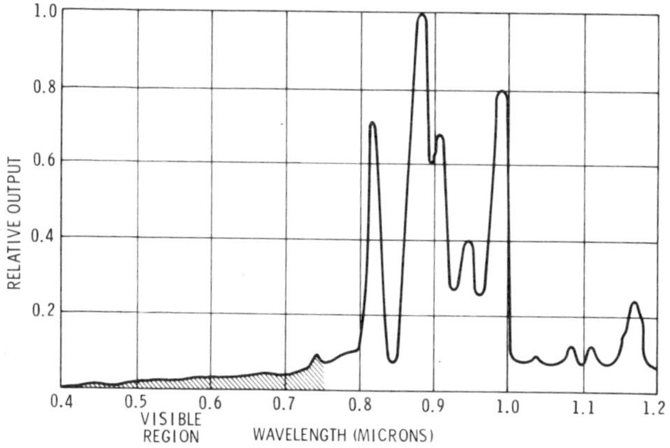

Fig. 1-3. Spectral output of 150-watt xenon arc lamp.

on a tungsten spectrum in Fig. 1-4 to illustrate the very-narrow-wavelength emission of these devices.

A typical LED is shown in Fig. 1-5. Though no longer manufactured, the silicon carbide diode in the figure is of interest since it is one of the first commercially produced visible-light emitting diodes. Like Round's crystal, the diode produces a yellow light. The construction of this and other kinds of LEDs will be described in this and later chapters; first, however, semiconductors and the mechanism of light emission in a semiconductor will be explained.

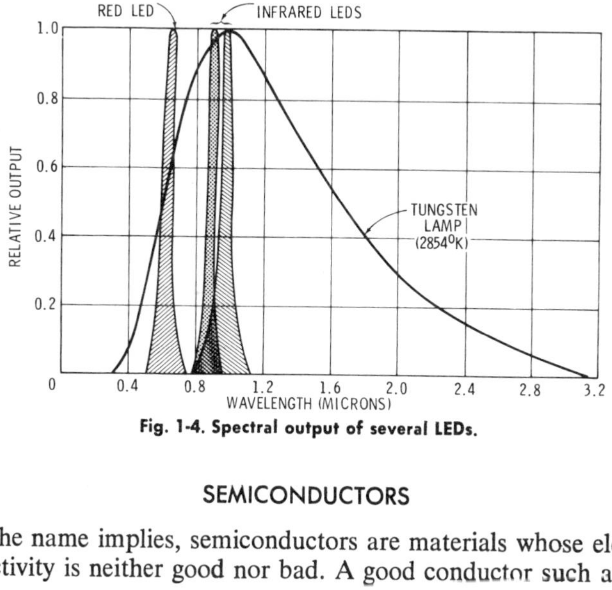

Fig. 1-4. Spectral output of several LEDs.

SEMICONDUCTORS

As the name implies, semiconductors are materials whose electrical conductivity is neither good nor bad. A good conductor such as silver

Courtesy General Electric Co.

Fig. 1-5. Silicon carbide SSL-1 LEDs and man-made silicon carbide crystals.

has a resistance of only 10^{-6} ohms/cm^3 while an insulator such as polyethylene may have a resistance of more than 10^{15} ohms/cm^3. Semiconductors such as germanium and silicon have resistances of 50 and 50,000 ohms/cm^3, respectively.

Electrical resistance is related to the number of electrons able to participate in conducting a charge from one point to another. The electrons that participate in carrying a charge are those in the outermost energy level of an atom. As shown in Fig. 1-6, an atom consists of a

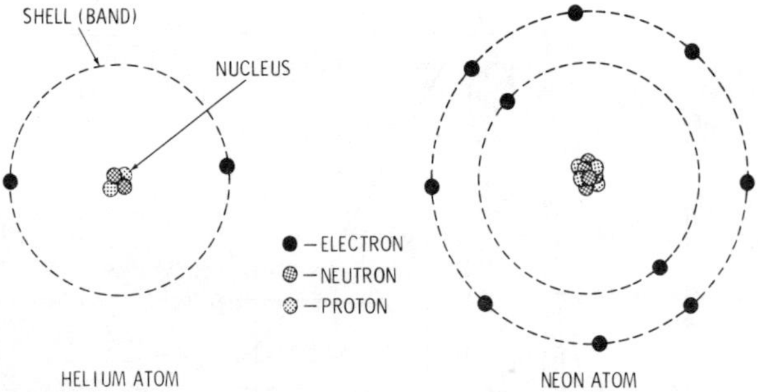

Fig. 1-6. Simple diagrams of two atoms.

nucleus containing positively charged protons, neutral neutrons, and various subatomic particles. The electrons in an atom occupy discrete levels or bands around this central core, with each level having a certain capacity for a maximum number of electrons. Beyond the first level, which holds only two electrons, the outermost band of electrons is full only when it contains eight electrons. This level is called the *valence band.*

The number of electrons in the valence band of an atom is representative of the element's stability. Atoms with only a few valence electrons are inherently unstable and readily combine with other atoms, but atoms with filled or nearly filled valence bands are very stable. For example, sodium has but one valence electron and can be explosively unstable, while xenon has a full complement of eight valence electrons and is, for practical purposes, chemically inert.

The valence electrons also determine an atom's electrical resistance. Insulators have five or more valence electrons while conductors have three or less. Semiconductors have four. Low valence electron counts give an atom freedom to temporarily give up an electron, and the fewer the electrons the stronger this tendency becomes. For this reason the very best conductors have but one valence electron. Conversely, the best insulators have a filled valence band unwilling to donate any electrons to the conduction process.

Pure semiconductors cannot be used in most electronic devices since their resistance is too high. The semiconductor must first be carefully contaminated with a small amount of impurity atoms. This procedure is called *doping,* and the impurity is called the *dopant.* Doping is usually accomplished during growth of the semiconductor crystal or by diffusion at elevated temperatures. It may seem more logical to employ a good conductor in the first place instead of contaminating a very pure, hence expensive, semiconductor. The reason for the use of the pure semiconductor is that by appropriate doping the material can be made to conduct electricity by either positive or negative charges. As we will see shortly, junctions of semiconductors doped in this fashion have the unique ability of permitting electrons to pass in but one direction.

Fig. 1-7 shows how a germanium crystal has been made to conduct by negative charges with the addition of a dopant such as phosphorus or arsenic. The germanium atoms have only four valence electrons, so each atom has sought the equilibrium of a full valence band by sharing its electrons with those of its neighbors in what is called a *covalent bond.* Phosphorus has five valence electrons, so a covalent bond of phosphorus and germanium will leave one leftover electron as shown in the figure. Since the dopant has contributed an excess electron, it is called a *donor.* This extra electron is free to move through the germanium crystal when acted upon by outside forces such as light, heat, or an electrostatic charge. Since the crystal has an excess of free electrons, it is called negative or n-type.

A semiconductor can be made to conduct by means of positive charges by adding a dopant with only three valence electrons such as indium or aluminum. The result is positive or p-type material. As shown in Fig. 1-8, the dopant combines covalently with the germanium host crystal to leave one unfilled slot. Since the dopant accepts an electron from the host, it is called an *acceptor*. The absence of an electron at each dopant atom with only three valence electrons is called a *hole*. Since the hole gives one of the germanium atoms less equilibrium than its neighbors, other electrons in the crystal can be drawn into the hole when they are liberated from their own atoms by light, heat, or other external forces. The end result is that the hole appears to have mobility, since it moves through the crystal as displaced electrons occupy its location and leave behind a hole of their own.

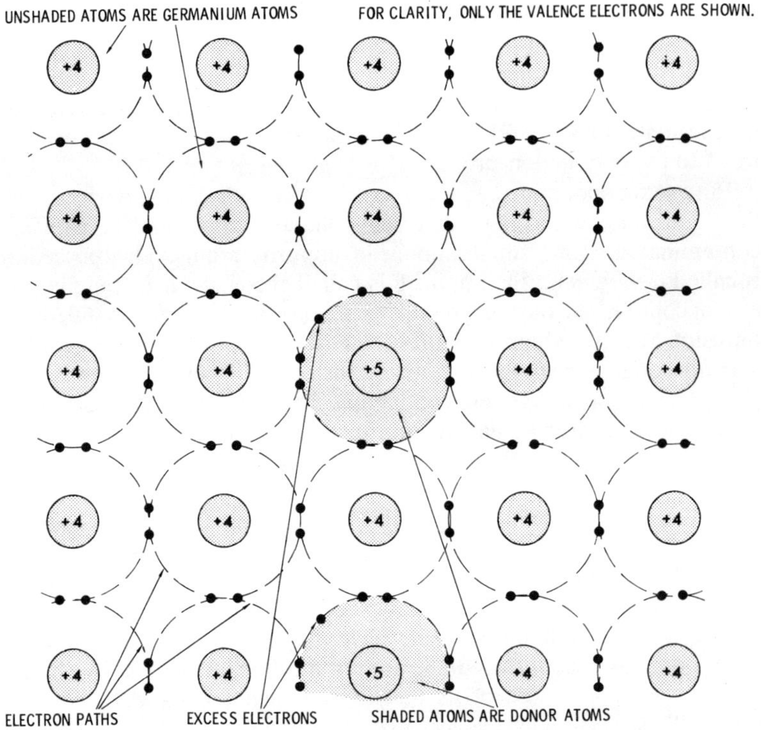

Fig. 1-7. Germanium crystal (n-type).

The movement of a nonexistent electron hole may seem difficult to comprehend at first. Actually it is perfectly all right to think of the electrons, and not the holes, as being mobile. But hole movement can frequently simplify semiconductor theory, so the concept is very commonly employed. The procedure is not much different from the movement of a bubble (hole) in a carpenter's level. When the level is inverted, the bubble is displaced upward, but at the same time the liquid (electron) is displaced downward. The end result is that both the bubble and the liquid have moved.

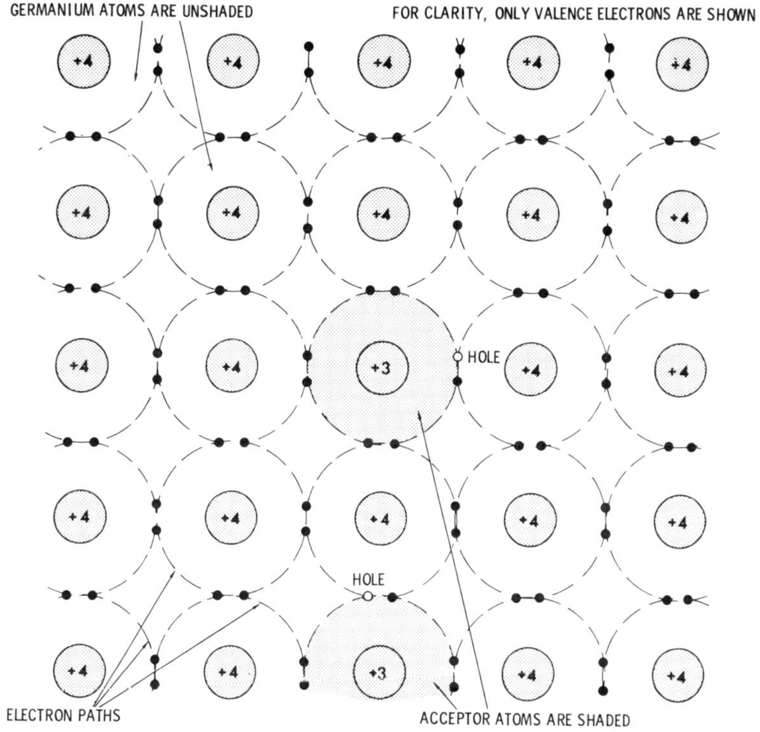

Fig. 1-8. Germanium crystal (p-type).

Summarizing, n-type semiconductors conduct an electrical current by means of free electrons while p-type material conducts with excess holes. The charge which carries an electrical current is termed the *majority carrier*. Therefore holes are majority carriers in p-type material and electrons are majority carriers in n-type material.

THE JUNCTION DIODE

By diffusing a p-type dopant into a wafer of n-type semiconductor as shown in Fig. 1-9, a pn junction can be created. Such a junction has the ability to conduct electricity well in one direction but poorly in the other. If one or more additional junctions are added, transistors, four-layer diodes, SCRs, and other devices with additional control over the current flow can be fabricated.

Diode action is based on the axiom that like charges repel while unlike charges attract. With no source of electrical current, some of the free holes and electrons in a thin layer along the pn junction recombine in an effort to establish equilibrium. The recombination region extends slightly beyond the junction itself and is called the *space charge region, depletion layer, potential barrier,* and other names.

When a negative charge is connected to the n side of the diode, the hole-rich region across the junction will attract the injected electrons. The electrons will readily move across the junction to drop into holes, and the result is a current flow. The diode is said to be *forward biased.*

If, however, the connections are reversed and the negative charge is connected to the p side of the diode, the electron-rich region across the junction will repel the injected electrons. The result is that few electrons cross the junction and there is negligible current flow. The diode is said to be *reverse biased.*

Both holes and electrons participate in current flow across a pn junction. While electrons cross the junction to fill holes, a hole "moves" across the junction to occupy spaces vacated by electrons. Like the bubble in the level, current flow in a diode can be described as either electron (liquid) or hole (bubble) flow.

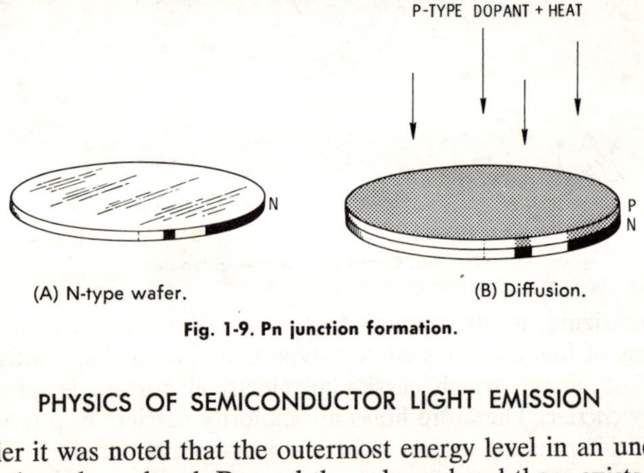

(A) N-type wafer. (B) Diffusion.

Fig. 1-9. Pn junction formation.

PHYSICS OF SEMICONDUCTOR LIGHT EMISSION

Earlier it was noted that the outermost energy level in an unexcited atom is the valence band. Beyond the valence band there exists a *conduction band* to which electrons can be elevated by an external force or stimulus. An atom with an electron in the conduction band is not in equilibrium, and there is a tendency for the stimulated electron to fall from the excited level back to the valence band (sometimes called the *ground state*). After a period called the lifetime, excited electrons spontaneously return to the valence band to recombine with the hole they previously vacated. In the process, they give off the energy absorbed during excitation as either light, heat, or both. The effect is summarized in Fig. 1-10. If the emission is predominantly light, the resumption of equilibrium is called *radiative recombination* and the emitted energy is called *recombination radiation.*

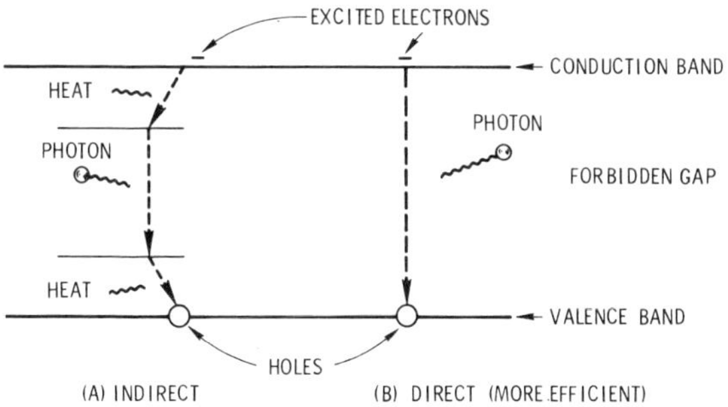

Fig. 1-10. Radiative recombination.

The wavelength of the radiation given off during recombination is related to the energy transition, and the formula that defines the wavelength is:

$$\lambda = \frac{hc}{E} \tag{1-1}$$

where,

h is Planck's constant (6.63×10^{-34} joule-seconds),
c is the velocity of light (3×10^{14} micrometers per second),
E is the energy in joules that separates the valence and conduction bands.

The equation can be simplified by converting to electron-volts. The result is

$$\lambda = \frac{1237 \text{ nanometers}}{E_g \text{ (electron-volts)}} \tag{1-2}$$

According to the equation, a material such as gallium arsenide with a bandgap separation of about 1.34 electron-volts would have a wavelength of 923 nanometers. Actually the wavelength is somewhat less (900 nm).

It should be noted that a well-defined bandgap would produce essentially monochromatic light, that is, radiation of one wavelength. By contrast, incandescent and fluorescent sources produce radiation of many simultaneous wavelengths.

Actually, nonlaser semiconductor light sources have a spectral bandwidth of typically 25 to 40 nanometers. This is much narrower than the emission of an incandescent source (see Fig. 1-4) but not nearly as monochromatic as most lasers. Nevertheless, since light emitting diodes have much narrower spectra than most conventional sources, they are frequently considered monochromatic.

The wide spectrum of a semiconductor emitter is due to variations in the bandgap and intermediate transition levels therein. Consider, for example, the semiconductor gallium phosphide (GaP). Zinc-doped GaP has a bandgap of 2.18 electron-volts and should therefore emit light at a wavelength of about 568 nanometers. In practice, GaP generally emits green light at a wavelength of 560 nm or more and if the concentration of phosphorus is sufficiently high, it will emit well into the red. Since the bandgap separation in even pure GaP is not fixed but may vary by several tens of nanometers, the output radiation is not monochromatic. A green diode, for example, may have an output that does indeed peak in the green at about 560 nm but that actually extends from below that wavelength through the yellow and orange and into the red. One side of the diode may emit in the green while the other emits in the red. Since the eye is more sensitive to green than red, the nongreen wavelengths are not dominant and the diode appears green.

The lifetime of a stimulated electron can range from a tiny fraction of a second to several days or more. For example, the atoms of some phosphors can be stimulated to high energy levels by means of ultraviolet light. When illuminated by infrared of the proper wavelength, the phosphor will experience sudden recombination, with the energy absorbed from the ultraviolet emitted as an orange glow. If no infrared is present, the phosphor will experience a gradual decay from the excited state to equilibrium over a *fluorescent lifetime* of a day or so. Most semiconductors have much shorter fluorescent lifetimes, with typical values being measured in microseconds or nanoseconds. For this reason, some semiconductor light sources can have exceptionally rapid response times.

Most materials are unacceptable from a light generation standpoint since their electrons do not fall from an excited state directly back to their normal level. Instead they may pause at intermediate levels along the way. The wavelength of electromagnetic radiation emitted by an electron is directly related to the energy level through which it falls; thus, a variety of spectral outputs will be had from these indirect bandgap materials, with few, if any, being in the visible region. Some materials, however, have the property of permitting excited electrons to fall from an excited to normal state with few, if any, pauses along the way. To fabricate an efficient solid-state light source, it is usually necessary to employ a material with such a direct bandgap structure. This is not always the case, however, and relatively efficient LEDs are made from indirect bandgap material.

Some semiconductors will emit light with little or no special preparation. Electroluminescent (EL) panels, for example, simply consist of powdered semiconductor pressed between two electrodes, one of which is transparent. Materials that readily produce light in this fashion include cadmium sulfide and zinc sulfide. Unfortunately these bulk-effect light generators are very inefficient. While EL panels have found some use in large-area light sources for aircraft cockpits, their inherent lack of efficiency and requirement for high operating voltage make them unsuited for many semiconductor applications. They do, however, have a relatively long life. An EL night light has been operating continuously in the author's home for more than three years.

LIGHT EMITTING DIODES

A simple way to efficiently generate light in many semiconductors is to fabricate the material into a pn junction diode. In this manner, stimulated electrons can be efficiently and directly used to generate light, and a great deal of control can be exercised over the fabrication of the light-emitting region and the efficiency of the device.

While the mechanism of light generation in an electroluminescent panel is not completely understood, there is a generally accepted theory for electroluminescence in a junction diode. As we have already noted, light generation in a semiconductor occurs when an excited electron resumes equilibrium by combining with a hole in the valence band. The junction diode provides a simple and efficient technique for placing large numbers of electrons in an excited state. In order for an electrical current to cross the potential barrier formed by the pn junction, electrons must be raised from the valence to the conduction band. After the junction has been crossed, the electrons fall into holes on the p side and in the process emit recombination radiation.

The light-emission process is summarized in Fig. 1-11 in a graph known as an *energy level diagram.* The valence and conduction bands on each side of the pn junction are connected by sloping lines representing the junction's barrier to electron flow. When electrons are injected into the n side by a source of current, the junction's potential barrier is reduced and current flow occurs.

The light emitted in such an arrangement is related to the bandgap of the semiconductor making up the diode. Most semiconductors require the controlled addition of dopant impurities to increase their electron hole concentration and mobility, and the result may be a variety of bandgap conditions that give rise to radiation of a lower energy (higher wavelength) than that of the original semiconductor. Depending on the design goals for a particular LED, this effect can be either highly desirable or disastrous. Semiconductors are far more transparent to radiation with a lower wavelength than that of the bandgap, so proper doping can significantly improve diode efficiency. LEDs, sometimes called solid-state lamps (SSLs), are distinguished by several important characteristics. Unlike most other light sources, they have response times ranging up to 100 megahertz or more and essentially monochromatic output. They can be operated continuously at temperatures ranging from that of liquid helium (near absolute

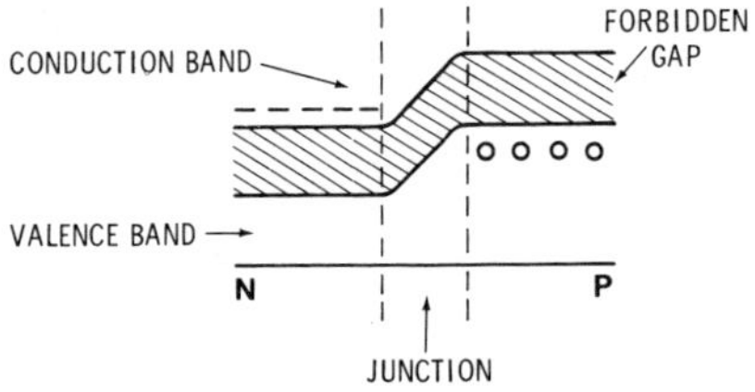

(A) No electron flow.

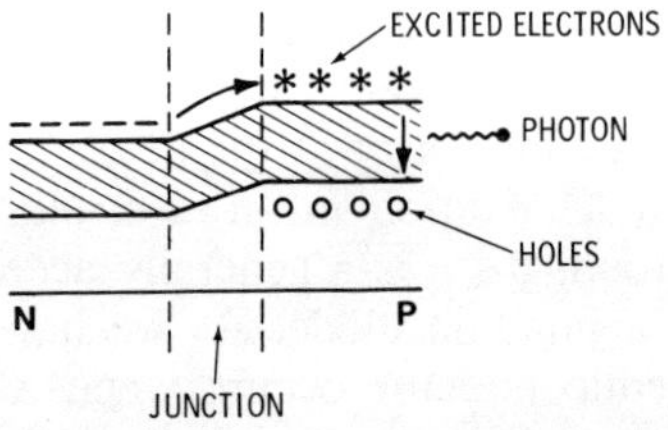

(B) Electron flow.

Fig. 1-11. Energy-level diagrams of a pn junction.

zero) to boiling water. The emission wavelength is reduced with temperature by about 2.5 angstroms (0.25 nanometers) per degree Celsius. At low temperatures, infrared LEDs may emit visible light and all LEDs are more efficient. And as was noted earlier, LEDs possess the reliability of a solid-state device.

Being a diode, an LED shows the characteristic voltage-current relationship of most pn junction diodes, i.e., current increases rapidly while voltage remains nearly constant, as shown in Fig. 1-12. The optical output for an LED is generally quite linear, with decrease in output due to overheating or to saturation of the light-emitting process. Fig. 1-13 shows a typical curve for power output as a function of current input.

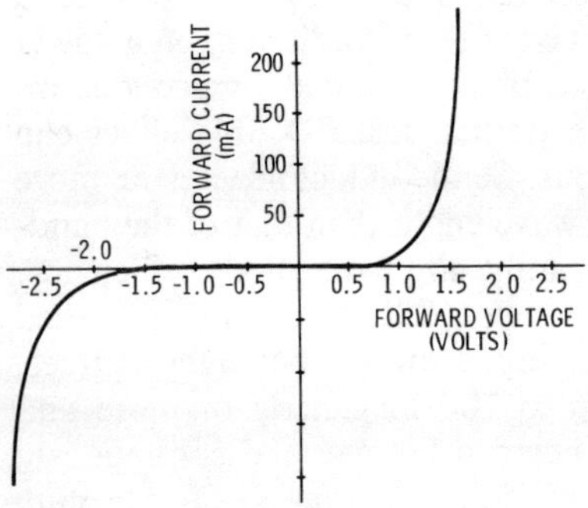

Fig. 1-12. Graph of voltage and current in an LED.

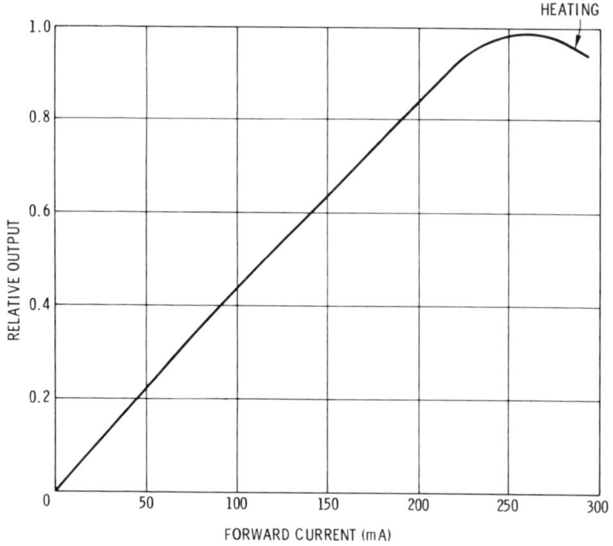

Fig. 1-13. Power output versus forward current in an LED.

VISIBLE AND INFRARED LEDs

By using semiconductors with various bandgaps, it is possible to obtain light emission from the blue portion of the visible spectrum to more than 30 microns in the infrared. Infrared diodes are well developed, and commercial units such as the one in Fig. 1-14 find use in optical communicators, intrusions alarms, and a host of other applications. Visible LEDs are used in great quantities as indicator lamps and numeric readouts. The calculator shown in Fig. 1-15 employs an array of such readouts.

Courtesy General Electric Co.

Fig. 1-14. SSL-5A gallium arsenide infrared LED.

Visible LEDs, readouts, and infrared emitters will all be described in great detail in coming chapters. Many of the manufacturing and operating techniques are similar or identical, so these topics will be discussed next.

Courtesy Hewlett-Packard

Fig. 1-15. HP-35 scientific pocket calculator with LED display.

LED MATERIALS AND FABRICATION

The most important step in manufacturing an LED is producing good quality semiconductor crystal. Crystal growing is the fine art of the semiconductor technology field, and there is strong competition among producers concerning whose material is the best.

A major requirement for semiconductor crystal is purity, but uniformity of structure is equally important. Polycrystalline structure is crystal containing a variety of internal orientations. The borders of adjacent planes can act as recombination centers for stimulated electrons, thus quenching light emission at a pn junction formed in the material.

A monocrystalline structure is sometimes called a single crystal since the entire crystal has a single crystalline structure and alignment. Monocrystalline material is hard to grow, but it is necessary for production of efficient LEDs.

There are excellent and detailed descriptions of semiconductor crystal growing in other references. Briefly, several techniques for forming single crystals are employed. The most common is the horizontal Bridgman technique, since it produces high-purity single-crystal material. In operation, the Bridgman method causes an ampoule of liquid gallium and arsenic to be moved through a high temperature gradient. As the molten semiconductor leaves the high-temperature zone, it solidifies or freezes into a crystal. If the ampoule is moved slowly and uniformly, subsequent solidification will be in the same plane as the original frozen material, and a large single crystal will be grown. Since impurities stay behind in the molten material, they are concentrated in the last portion of the ampoule to pass through the temperature zone. Therefore, the majority of the crystal is of high purity.

A Bridgman crystal is formed in the shape of the ampoule or "boat" in which the original material was placed and therefore has a very uniform shape. The crystal, sometimes called a *boule,* is tested for orientation of its crystalline structure and then sliced into very thin wafers with a diamond saw or a wire saw. LEDs are formed from the wafers.

Two techniques can be used to form a pn junction in each of the wafers. The simplest is diffusion, a process in which the wafer is heated in the presence of an acceptor dopant such as zinc. The diffusion temperature is usually in the range of 800 to 900°C and the time can be up to several hours or more. The diffusion depth, and hence the position of the pn junction, is increased for long diffusion times. After diffusion, the wafer is lapped, polished, and cleaned. Sometimes a metal-contacting layer is plated or evaporated onto one or both sides of the wafer. It is then divided into hundreds of small chips about 20 or 25 mils square, which form the individual LEDs. The process is summarized in Fig. 1-16, and a typical diffusion furnace is shown in Fig. 1-17.

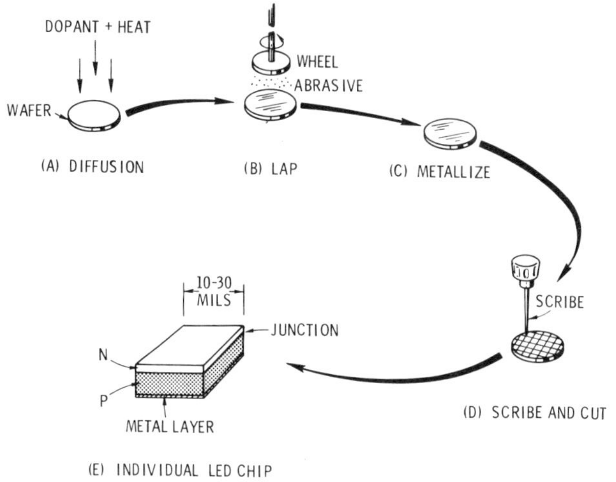

Fig. 1-16. Typical LED manufacturing process.

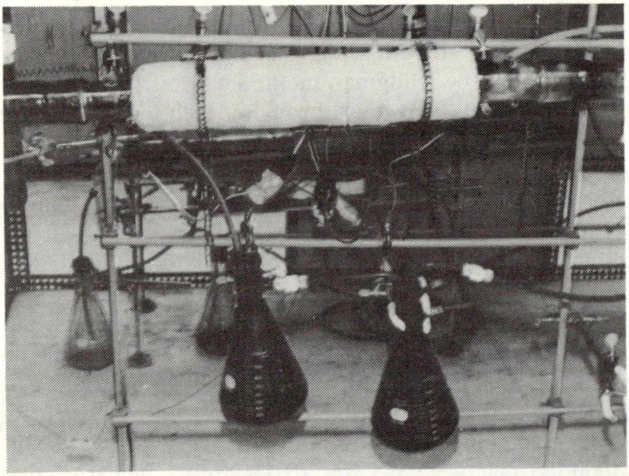

Fig. 1-17. Spectronics LED diffusion furnace.

Low-cost visible LEDs are frequently made with individual contact regions and junctions by means of the same photolithographic technique used to make transistors and integrated circuits. In this technique, a wafer is coated with a material such as silicon dioxide, and an array of small apertures is etched into the coating. The entire wafer is then diffused to simultaneously form hundreds of individual junctions. Metal contacts are made by applying silicon dioxide to the exposed light-emitting region and etching away areas to be coated. Finally, the wafer is cut up into individual diodes ready to be mounted.

LED chips are packaged in a variety of techniques, several of which are described later in this chapter. Almost always the absorbing p side of a chip is soldered to a large-area contact, while a small electrode wire is bonded to the low absorbing n side. In this way more of the LED's radiation is permitted to escape.

In recent years a newer technique of junction formation has become the preferred method for making some LEDs. There are several variations, but the most common is called *liquid phase epitaxy*. In this technique a wafer of n-type material is mounted at one end of a boat-like crucible and inserted into a furnace capable of being tilted from the normal horizontal position. The furnace is tilted to about 30° from the horizontal and a quantity of gallium, gallium arsenide, and a p-type dopant such as silicon is added to the portion of the boat away from the wafer. The materials vary according to composition of the substrate wafer. The furnace is then heated for about an hour at 850°C in order to properly dissolve the gallium, gallium arsenide, and dopant into a saturated mix. Then the furnace is tilted past the horizontal position to 30° in the opposite direction to cause the molten mix to flow over

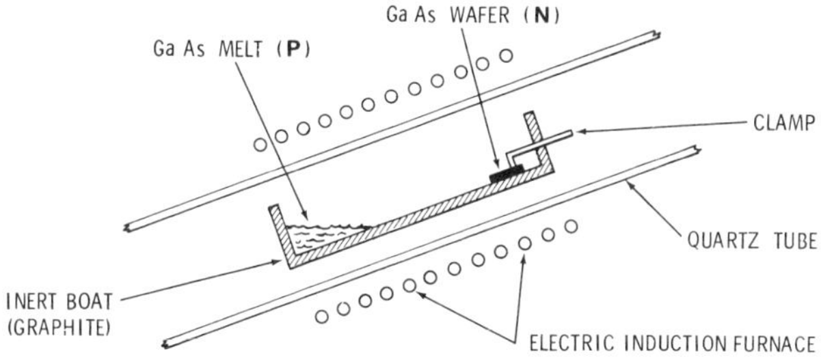

Fig. 1-18. Epitaxial furnace.

the wafer. The furnace temperature is raised to about 900°C, and a very thin layer of the wafer melts and dissolves in the molten mix. Next the temperature is lowered to about 700°C and the mix begins to deposit a thin layer of crystal onto the freshly exposed wafer surface. The result is a thin epitaxial layer of very uniform structure with the same alignment as the host wafer. Fig. 1-18 shows the furnace and boat used to form the epitaxial junction.

The epitaxial junction is characterized by a high degree of uniformity and is therefore preferred for the production of high-efficiency light emitters. But the diffusion process has the advantage of permitting many individual diodes to be formed at once by means of photolithographic techniques. Therefore, both junction-forming methods find extensive use in the production of commercial LEDs.

With either technique of junction formation, individual chips may produce more or less optical power than others. This is due to a variety of factors, including impurity centers, junction uniformity, crystal dislocations, and contact resistance, Since diodes produce a range of powers, they must be graded with a calibrated detector before being

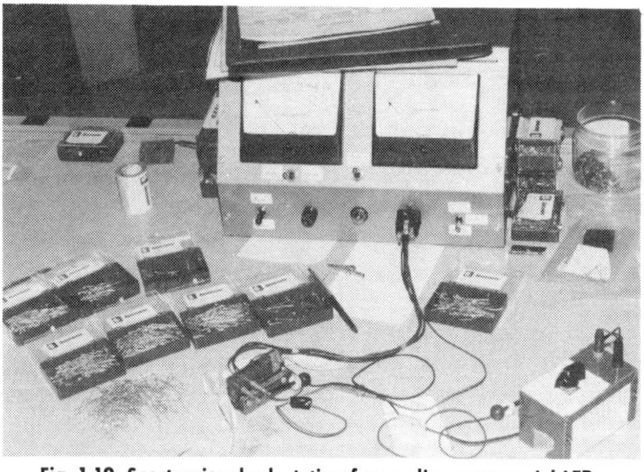
Fig. 1-19. Spectronics check station for grading commercial LEDs.

sold commercially. Fig. 1-19 shows a check station at Supertronics, Inc., for sorting diodes in this manner.

LED STRUCTURES

Several different structures designed to improve the optical efficiency of an LED will be mentioned in the next section. These include specialized structures and conventional geometries designed for low-cost mass production. To provide a better understanding of the most significant structures, several are described and illustrated here.

Flat Emitter

Flat LEDs employ a variety of pn junctions and contacting techniques. Surface emitters generally have a small electrode on the n side and a large-area contact on the p side. Light is emitted from all exposed parts of the diode. The junction may form a plane through the diode as in the case of epitaxy or diffusion, or it may be isolated to a central region by masking and diffusion. The latter technique, a planar junction, has the advantage of isolation of the junction region by the surrounding semiconductor.

Edge emitters employ evaporated-metal contacts on both sides of the diode. While this improves power efficiency by reducing contact resistance, light is emitted only along the plane of the junction.

Fig. 1-20 shows both surface- and edge-emitting flat LEDs. Both configurations find wide use in commercial LEDs. Frequently, edge emitters employ a small parabolic reflector to collect the LED radiation and project it into a beam. Some diodes employ the reflector to collect the edge emission from a surface emitter as well.

Mesa Emitters

The mesa emitter is actually a variation of the flat emitter. Fig. 1-21 shows a mesa structure made by Ferranti and a more conventional mesa LED. The Ferranti structure is unique in that the LED is inverted and the radiation emitted from the n side of the junction is reflected outward by the shiny header. Because they are more difficult

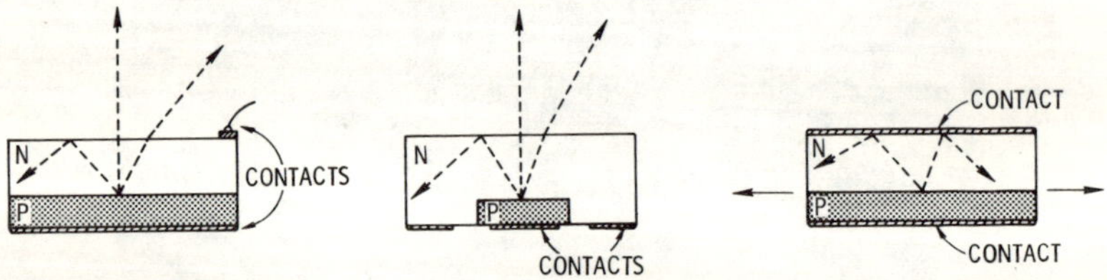

(A) Surface emitter. (B) Planar surface emitter. (C) Edge emitter.

Fig. 1-20. Flat LED configurations.

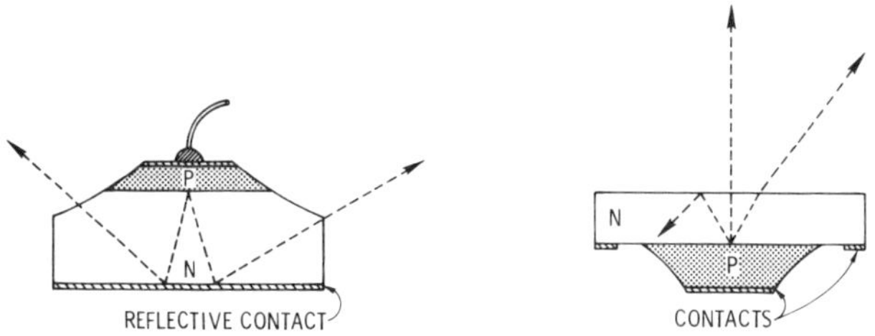

(A) Inverted mesa.

(B) Conventional mesa.

Fig. 1-21. Two mesa-structure LEDs.

to make than flat emitters, mesa structures are not frequently employed in commercial LEDs.

Dome Emitters

The most unusual LED structure is the hemispherical dome. As shown in Fig. 1-22, the dome structure consists of a conventional pn junction formed into the back side of a semiconductor dome. The dome increases light-emitting efficiency by reducing the reflection which occurs at the semiconductor/air interface of flat diodes.

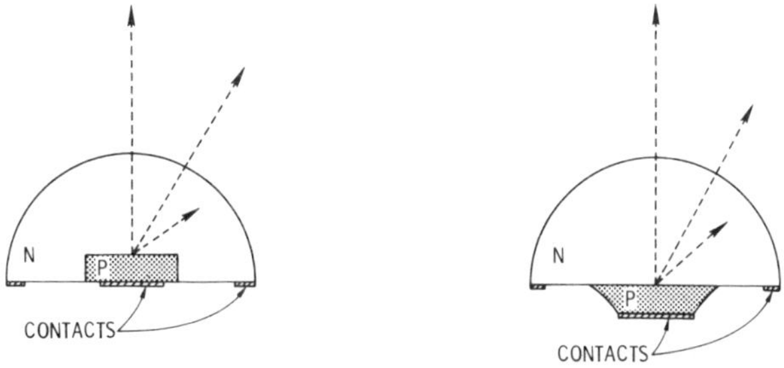

(A) Planar.

(B) Mesa.

Fig. 1-22. Two dome-structure LEDs.

The literature describes two variations on the dome structure, planar and mesa. The planar dome employs an immersed pn junction formed by masking and diffusion while the mesa dome employs an exposed pn junction. The mesa diode can be made by forming a diffused or epitaxial layer of p-type material at the back side of the dome and then etching away all but a central, raised area.

While domed emitters are very efficient, they are also very expensive. They use much more semiconductor crystal than other LEDs, and the domes are difficult to manufacture and mount. The domes are made by cutting cylinders from a semiconductor wafer and grinding them into hemispheres. Mounting is made difficult by the need for aligning the central p region and the annular n region contacts with their respective contact points on the mounting header.

Other Emitters

Numerous variations of the three basic structures are possible. For example, dome emitters can be formed into other shapes designed to reduce the internal reflection phenomenon. Flat structures can be modified into bars for LED numeric displays.

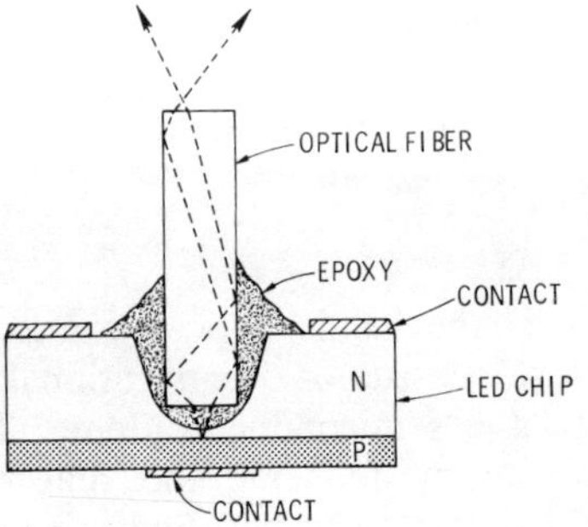

Fig. 1-23. Bell Telephone Laboratories LED with optical fiber.

One interesting modification is a diode designed by two Bell Telephone Laboratories scientists. Shown in Fig. 1-23, the diode incorporates a planar diffused junction and an etched depression immediately over the junction region. The depression is designed to allow an optical fiber to be epoxied as close as possible to the active light-emitting region in order to provide good optical coupling.

LED EFFICIENCY

In an ideal diode light source, each incoming electron would result in the emission of a photon. This would give an *internal quantum efficiency* of 100 percent. In practice, imperfections within the semiconductor may give rise to recombination at crystalline defects or regions of impurity concentration and resultant generation of heat instead of light.

Even assuming a light emitting diode with a perfect quantum efficiency, many conditions can exist which prevent internally generated photons from being emitted by the device. Frequently the p side of the junction is more highly absorbent than the n side, though emitted photons can be trapped by both sides. The contacts necessary to provide electrical power can block much of the radiation, particularly if one side of the diode chip is mounted to a metal tab whose surface area equals or exceeds that of the chip. And even those photons which survive the trip to an open area on the LED surface may be reflected back into the diode because of the index of refraction mismatch between the semiconductor and air.

As a result of these and other factors, a diode with a high internal quantum efficiency can have a relatively low *external quantum efficiency*. For example, in 1965 William N. Carr of Texas Instruments, Inc., measured an internal quantum efficiency of 88 percent in specially prepared GaAs LEDs. Unfortunately, these same diodes exhibited an external quantum efficiency of only 42 percent (which nevertheless is still quite good).

Carr found that the loss in external efficiency over that theoretically possible was due to the absorption of light within the crystal making up the diode and the blocking of light by the header upon which the diode was mounted.

Another figure of merit for LED operation is *power efficiency*. Due to resistance at the contact regions, an LED will not receive all the electrical current supplied by a power source. Resistive loss can also occur within the semiconductor itself. The result is that the power efficiency of an LED is never 100 percent. Power efficiency is dependent on resistive losses and external quantum efficiency, and it is typically a few percent. Some specially fabricated diodes may have a very high power efficiency of 10 percent or more.

IMPROVING LED EFFICIENCY

A variety of interesting techniques have been devised in order to extract as much light as possible from an LED. To avoid absorption within the semiconductor making up the LED, diodes are fabricated so that the light generated at the junction need travel but a short distance before being emitted into air. Since the n side of an LED junction absorbs much less light than the p side, the n side is usually employed as the main light-emitting region.

Several new techniques for making contact to LEDs have been developed to preclude excessive blocking of internally generated light by metal contacts. Some techniques employ circular electrodes, while others use a single dot contact. Another technique is to ignore radiation emerging from the top of the LED and collect only that emitted from the edge of the diode in the plane of the junction. In this manner large-area contacts can be used on the top and bottom sides of the LED, with a resultant increase in electrical work function. This technique can greatly reduce the series resistance of an LED and improve the power efficiency of the device.

Since half the light generated at an LED junction will travel toward the header upon which the chip is mounted, the potential external efficiency of the diode is significantly impaired. Some manufacturers have employed reflective headers to cause light which would otherwise be wasted to be reflected back into the diode structure. Since the photons striking the reflective header and bouncing back through the diode must pass through the lossy p region twice, many of them are absorbed. Some are also absorbed by imperfections in the reflective header, but enough complete the double pass through the LED structure to improve external quantum efficiency.

A major problem in extracting light from an LED is caused by the high index of refraction of most semiconductors. Refraction is caused by the difference of the speed of light in air and a particular substance and is given by:

$$n = \frac{c}{V} \tag{1-3}$$

where,

c is the velocity of light in air,

V is the velocity of light in the material whose refractive index is being measured.

Air has an index of refraction of 1(1.0003 when compared to a vacuum) while other materials are higher. The index of refraction of diamond is 2.42, and gallium arsenide, with a value of about 3.5, is one of the few materials that exceed this value. This means that light striking a GaAs surface at an angle *greater* than about 16° is totally reflected back into the crystal. The angle at which light striking the surface is totally internally reflected is called the *critical angle,* and several interesting techniques are employed to increase it.

One method is to roughen the surface of the LED to improve the chances for photons striking the semiconductor/air interface within the acceptance angle. However, this technique can be impractical. A better method is to cover the light-emitting portions of the LED with an antireflective coating. The coating permits light that arrives at an angle greater than the critical angle of the semiconductor to be emitted. However, the coating is fragile and can be difficult to apply when contacts are present. For the coating to be effective, it must be applied in a carefully controlled thickness, and it may have a tendency to peel off if the LED is not properly cleaned before application. A GaAs LED coated with a silicon oxide (SiO) film having a refractive index of 1.9 showed an improvement of 35 percent over a similar uncoated unit. Theoretically the SiO coating should permit up to a 45-percent improvement in output.

The third method of increasing the critical angle is to alter the shape of the semiconductor from which the LED is made. This is the technique used in the high-efficiency diodes tested by Carr at Texas Instruments. These diodes were formed into a hemisphere by an ultrasonic cutting and abrasive grinding process. Since the surface of the diode is curved, light originating at the junction always arrives at an angle within the critical 16°. Diodes can be formed into a hemisphere, truncated sphere, truncated cone, truncated ellipsoid, paraboloid, and other shapes, using grinding or other esoteric techniques. While all these fabrication methods greatly increase external quantum efficiency, they are very expensive. Hemispherical diodes commonly sell for more than a hundred dollars in single quantities.

It is important to note that while a hemispherical LED can theoretically improve external efficiency over an equivalent flat device by a factor of 26, bulk absorption within the thick n region (required for the hemisphere) causes losses that restrict actual improvement to about 10 times. Also, the absorption characteristics of the n region may slightly increase the peak wavelength of the emitted radiation. Nevertheless, domed diodes are the most efficient LEDs available, with power efficiencies ranging to 10 percent and higher.

In order to duplicate the advantages of the expensive LEDs with curved surfaces, manufacturers have learned to encapsulate their devices in a variety of transparent plastics. Those plastics currently employed have an index of refraction ranging from 1.4 to 1.8, certainly not a match for the 3.3 to 3.6 value of light-emitting semiconductors. Nevertheless, the improvement is substantial. In a typical GaP LED, the critical angle is 17.7°, but addition of a transparent epoxy with an index of refraction of 1.66 increases the angle to 30.3°. The result is 2 1/2 times more light output.

Several manufacturers are experimenting with various kinds of arsenic-chalogen-halogen glasses that have refractive indexes ranging from 2.4 to 2.9. An infrared-emitting diode made of GaAlAs and coated with such a glass demonstrated an external efficiency of 7.2 percent, an improvement of 4 to 6 times over a similar uncoated unit. The glass, of course, has a much higher melting point than plastic and is therefore more difficult to apply. Epoxies have the advantage of room-temperature application and subsequent hardening.

Those various techniques of improving the external efficiency of LEDs have a variety of advantages and disadvantages. Some were discussed in this section, and others are discussed elsewhere in this book. But for each technique to be effective, the mechanical structure upon which the LED is mounted must offer as little resistance to emitted light as possible. This is the packaging problem.

PACKAGING

Like any other semiconductor, an LED must be encased in a protective package. The package provides mechanical rigidity for the device, keeps away moisture, dirt, and other contaminants, and holds the delicate connecting lead or leads in place. Additionally, the package serves to provide external electrical and mechanical contact to the

LED. Frequently the package provides a mount for an optical device such as a lens to enhance the emission efficiency of the diode. In most low-cost LEDs an epoxy blob provides both the protective package and the optical element.

Most early LEDs were mounted on conventional transistor headers and encased in glass-windowed cans because of their ready availability, and even now many LEDs are packaged in this fashion. Fig. 1-24 shows a typical LED metal header and glass-windowed protective can before assembly. The can is welded to the header to provide a hermetic

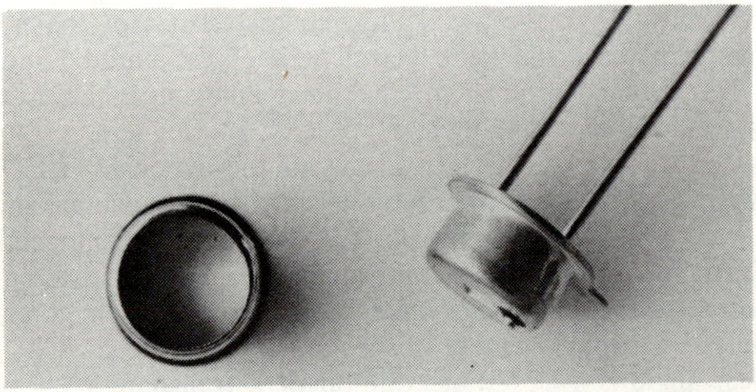

Fig. 1-24. Header and glass-windowed protective can for LED.

seal. Unfortunately the can blocks much of the LED's radiation and is therefore not an optically efficient packaging technique. Plastic is not nearly as sturdy as metal, and therefore some degree of optical loss can be tolerated in applications requiring a well-protected LED.

Several mounting techniques are shown in Fig. 1-25. Subsequent chapters include more on packaging, and Chapter 5 describes various methods to efficiently collect light from each of the common packaging techniques by means of external optics.

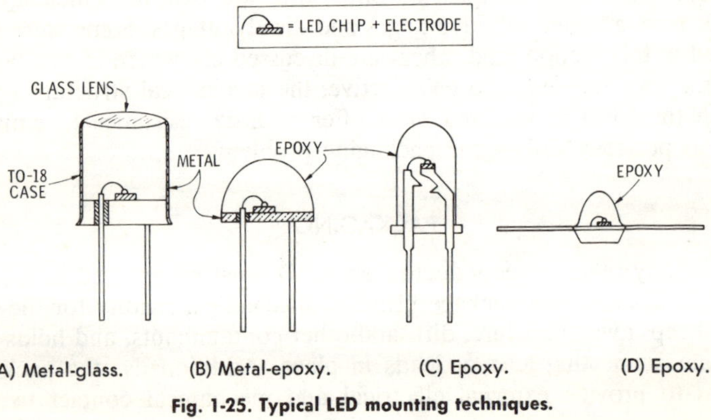

Fig. 1-25. Typical LED mounting techniques.

RELIABILITY

With their virtually unlimited operating life, semiconductor LEDs possess a singular advantage over all other artificial light sources. When operated at rated current levels a typical LED can be expected to emit more than half its maximum power rating after many years of continuous operation. Some manufacturers talk in terms of 100,000- or 200,000-hour half lives. Since a year has only 8760 hours, this is from 11 to 22 years. Even after losing half of its light output, a visible LED may seem almost as bright to the human eye as when the device was operating at full power. Therefore LED indicator lamps have excellent potential for widespread use, particularly since miniature incandescent lamps designed to operate only 10,000 hours are far more costly.

Even though LEDs can have very long operating lifetimes, there is a degradation mechanism that eventually reduces the power output of a diode. The mechanism is dependent on current density, and diodes operated at low current levels have significantly longer lives than those operated at maximum allowable levels.

LEDs can also be immune to environmental stresses, but this largely depends on the packaging technique. Texas Instruments found that the comprehensive tests required for the tiny TIL23/24 series to pass military standards resulted in only one failure in several hundred diodes. All of the diodes tested survived temperature cycling, thermal shock, moisture resistance, mechanical shock, acceleration, and vibration tests, and only 1 of 390 failed the hermetic seal test.

The TIL23/24 is housed in a metal and ceramic pill package with an integral glass lens, and diodes without the protective advantages of such robust packaging are not nearly as reliable under environmental extremes. For applications where there is a need, the additional cost of a sturdy mounting package can be well justified by the added operating life.

REDUCING DEGRADATION

A fascinating paper by two Bell Telephone Laboratories scientists several years ago described a unique way to extend the life of an LED. The paper was written by C. A. Burrus and R. W. Dawson and appeared in *Applied Physics Letters,* August 1, 1970 (pp. 97-99).

In their paper, Burrus and Dawson described a technique for coupling the optical output of an LED to a small optical fiber for a higher radiance output. The fiber was epoxied to a depression etched in the n side of the GaAs diode, as shown earlier in Fig. 1-23. The unique aspect of the paper was the discovery that degradation of the diode could be prevented by applying an occasional reverse bias to the LED.

The reverse bias could take the form of a 0.02-volt or greater (to the breakdown point) negative bias applied at duty cycles of from 1 to 10 percent. The results of the reverse-bias pulsing were dramatic, with diodes tested showing significantly reduced degradation. Some diodes even experienced a substantial *increase* in power output as a result of the negative biasing. Fig. 1-26 shows some of the results of the test.

Burrus and Dawson also found that a previously degraded diode could be brought back to its original operating power output by simply heating the device for from several hours to a few days at 100 to 200°C. An external reverse bias could be applied if desired; however, the internal electrostatic bias of the diode could also be used. The diode could be rejuvenated again and again with the simple heating technique.

Both techniques described in the paper can be applied to many commercial LEDs. The reverse-bias technique is particularly easy to use, and some circuits can be easily modified to produce a small-amplitude negative pulse following the larger driving pulse. The heating technique to restore a previously degraded diode could even be used with epoxy diodes so long as the maximum temperature rating of the device, usually at or near 100°C, is not exceeded.

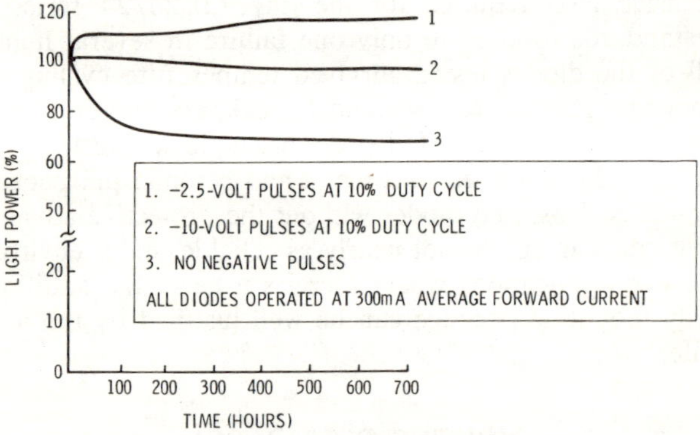

Fig. 1-26. Reduction of LED degradation by reverse pulsing.

SEMICONDUCTOR DIODE LASERS

A particularly unique and important light emitting diode is the semiconductor injection laser. By modifying a conventional flat GaAs LED so that two opposite ends perpendicular to the plane of the junction are perfectly flat and parallel, the diode can be made to act as a laser. An important prerequisite for laser action is a very flat, uniform junction region. By injecting very brief pulses of high-current electricity through the diode, a condition will result where more electrons are in an excited state than in an unexcited state. As some of the electrons recombine with holes, their recombination radiation stimulates the recombination of other electrons and holes and the emission of additional photons. This process is called *stimulated emission of radiation* and is essential for the onset of laser action.

In a light emitting diode, photons are generated randomly as electrons and holes spontaneously recombine. For this reason LEDs are sometimes called spontaneous emitters. At very high current levels some stimulated emission can occur in an LED when a few of the photons stimulate recombination, but the process is very inefficient, with only a small number of photons taking part in it. The two end mirrors on the semiconductor diode laser set up a condition that encourages far more stimulated emission than in an LED.

The laser action itself begins in earnest when some of the recombination radiation strikes one of the two end mirrors and is reflected back into the junction region. Since the reflected photons continue to stimulate recombination radiation during their passage between the end mirrors, an avalanche of photon generation is created. The end mirrors, which are almost always formed by cleaving the GaAs material along parallel crystalline planes, have a reflectance of about 35 percent, so some of the internally generated light is emitted from each mirror on each pass. Frequently a thin layer of gold is evaporated over an insulating film of SiO on one of the mirrors to give almost 100-percent reflectance. In this manner all of the light can be caused to exit from one end of the laser.

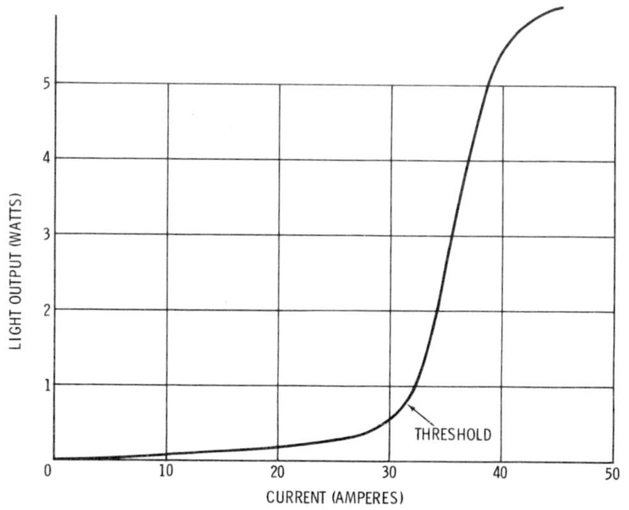

Fig. 1-27. Graph of laser current versus laser optical output.

Below a certain threshold level the injection laser produces incoherent light, and its characteristics are virtually identical to those of a conventional LED. When the population inversion is attained by means of high-current driving pulses, however, the laser diode demonstrates a dramatic nonlinear increase in light output (as shown in Fig. 1-27) and a narrowing of spectral width. While a conventional GaAs LED might have a spectral width of several tens of nanometers, the output of a GaAs laser may be only a few. Another characteristic which demonstrates that the device is behaving as a laser is the directional nature of the light emitted from the end mirrors. An LED may have a very large beam pattern, but radiation from an injection laser is contained in a definite beam with a divergence of 15° to 30°. This effect is shown in Fig. 1-28.

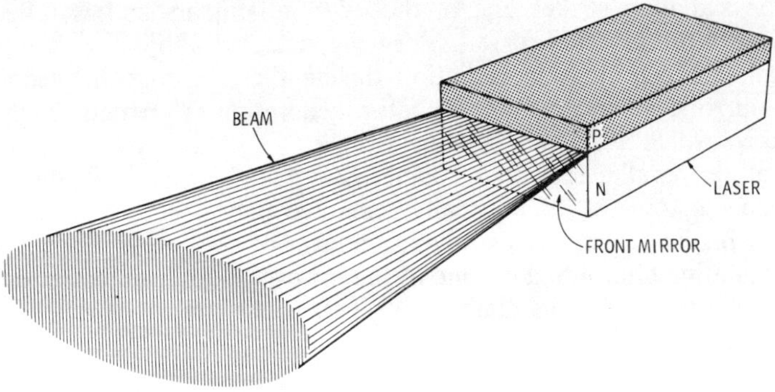

Fig. 1-28. Injection laser beam.

LASER MATERIALS AND FABRICATION

Since the first injection lasers were invented almost jointly by General Electric, IBM, and MIT in 1962, light emitting diodes made of several dozen semiconductor combinations have been made to lase. Output wavelengths range from more than 20 microns in the infrared for lead salt devices to visible red light for aluminum- and phosphorous-doped GaAs units. The best-studied and least-expensive lasers are made with GaAs and emit in the near infrared at about 910 nm at room temperature. As with LEDs, the wavelength of an injection laser shifts downward as temperature is decreased. When immersed in liquid nitrogen, for example, a GaAs laser or LED will emit at about 845 nm.

The fabrication process for injection lasers is in many respects similar to that of LEDs. In both cases high-quality, defect-free semiconductor material is required, and careful attention to materials processing and cleanliness is a must. A typical GaAs injection laser is made by forming a flat, uniform junction in a GaAs substrate via diffusion or epitaxial growth. The crystal wafer is then lapped to a uniform surface texture and cleaved into thin bars with parallel, reflecting sides. Cleaving is along crystalline planes and is frequently accomplished by scribing with a sharp tool.

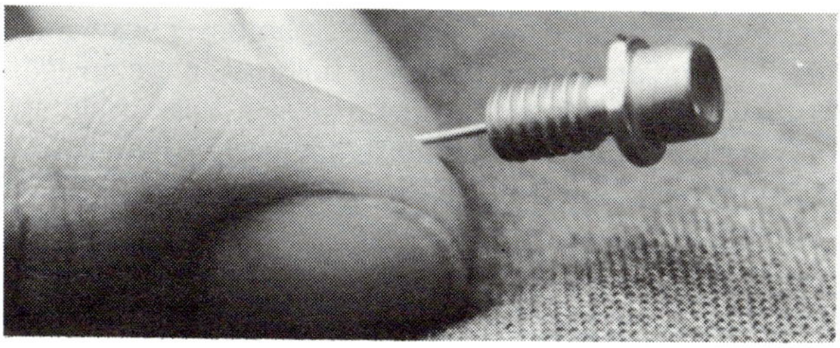

Fig. 1-29. RCA injection laser diode.

Individual lasers are formed by sawing the bars into dice. The sawing, which is usually accomplished with a wire saw, roughens the sides of the individual lasers and prevents undesirable "cross-lasing" which would result if all four sides formed two parallel reflecting pairs. Finally, the lasers are soldered to headers and protected with a plastic cover or glass-windowed can. A low-cost diode laser is shown in Fig. 1-29.

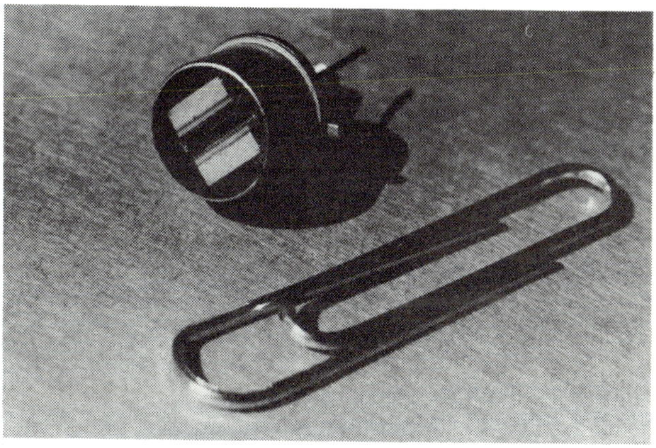

Fig. 1-30. RCA injection laser diode array (64 individual lasers).

For more power, laser diodes are mounted in arrays or stacked on top of one another. While a stack may have only two or three lasers, an array may incorporate thousands of individual laser chips capable of producing a beam with a peak power in excess of a thousand watts. Smaller arrays, such as the unit shown in Fig. 1-30, may produce a hundred watts or more.

LASER OPERATION

Conventional injection lasers require very high current levels to achieve a population inversion. Since the series resistance of a laser can be as high as ½ ohm or more as a result of inefficiency of the laser in converting injected electrons to emitted photons, considerable heating is caused by the driving current. For this reason a conventional injection laser will be destroyed if operated at a duty cycle exceeding about 0.1 percent at room temperature. Typically, individual current pulses across the laser must not exceed a width of 200 nanoseconds, for safe operation.

These operating restrictions complicate pulser design and require the use of fast rise time photodetectors and amplifiers. However, if a few basic precautions are followed, it is not difficult to assemble suitable pulser circuits. Chapter 2 describes several high-current pulsers suitable for driving both injection lasers and LEDs.

LASERS vs LEDs

An important consideration in the design of some kinds of electro-optical equipment is whether to employ a laser diode or LED. Some tradeoffs are obvious. For example, LEDs are superior to lasers as indicator lamps, short-range detection systems, electro-optic couplers, film annotators, and dozens of other applications where either visible or infrared light is required at low-to-moderate power levels. The question becomes more complex at higher power levels.

Let us compare a commercial injection laser capable of delivering 4.5 watts when driven with a 10-ampere pulse (such as a selected RCA SG2002) with an efficient and inexpensive LED capable of delivering 6 milliwatts at 100 milliamperes (such as the GE SSL-55C). When driven with a 10-ampere pulse, the SSL-55C will emit 240 milliwatts, only about 5.3 percent of the 4.5 watts delivered by the laser for the same peak current. Furthermore, at this high current level the power efficiency of the LED has dropped from 4.5 to 5 percent at the 100-mA level to only 1.7 percent, while the power efficiency of the laser is an impressive 5.6 percent.

So far the laser is ahead, but there are other points to consider. The LED can withstand pulses up to 2 microseconds in duration at 10 amperes, while the laser is restricted to a maximum width of 0.2 μsec (200 ns). Also, the laser costs considerably more than the LED and consumes more power (80 watts vs 13.5 watts for the LED). Another important factor concerns degradation. While there have been important advances in injection laser technology in recent years, laser diodes are known to suffer from internal degradation. After a few thousand hours of operation, a "good" laser may deliver up to 80 percent of its

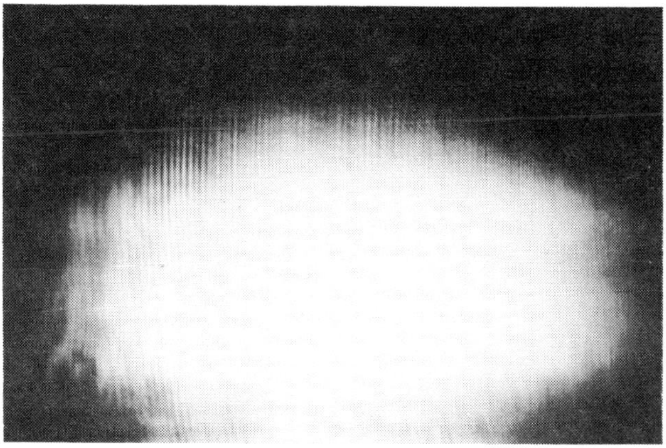

Fig. 1-31. Injection laser beam pattern on infrared film.

rated output when new. LEDs also suffer from degradation, but the effect is generally much less pronounced.

Another point to consider is the inhomogeneities of a diode laser beam versus the very uniform beam from most LEDs. Also, the laser is easily collimated because of its very small source size. A typical laser beam versus the very uniform beam from most LEDs. Also, the laser red film and shows the complex structure within the laser beam.

Many of these tradeoffs are summarized in Table 1-1, and subsequent chapters will again compare the relative merits of injection lasers and LEDs.

ROOM-TEMPERATURE LASERS

Until 1970 all semiconductor lasers had to be operated in a pulse mode at room temperature, since the high current density required to reach threshold caused significant heating. Continuous operation could only be achieved at very low temperatures. In 1970, scientists at Bell

Table 1-1. LEDs vs Lasers

Characteristic	LED	Laser
Peak power	300 mW	100 W
Average power	250 mW	15 mW
Power efficiency	10%	6%
Maximum duty cycle	100%	0.1%
Rise time	1–300ns	1 ns
Beam divergence	10–180°	20°
Spectral width	40 nm	3.5 nm

Note: These are representative optimum values.

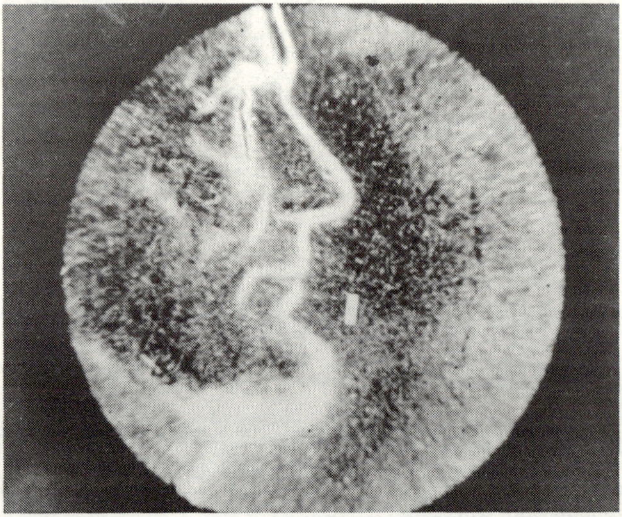

Courtesy Bell Telephone Labs

Fig. 1-32. DH semiconductor laser chip on a Lincoln penny.

Telephone Laboratories developed a laser that has a threshold so low that the device can be operated continuously at room temperature. The laser is shown on the head of a Lincoln penny in Fig. 1-32. The device has electrical characteristics very similar to those of an LED.

The new continuously operating lasers employ an optical waveguide to reduce threshold. The waveguide is formed by sandwiching a thin GaAs layer between two gallium aluminum arsenide (GaAlAs) regions. Recombination takes place at the junction of the GaAs and one of the GaAlAs layers; the emitted radiation is kept in the active region by means of the reflective nature of the interface of the two regions. The reflection is caused by the index of refraction mismatch between GaAs and GaAlAs. Because of the construction of the diode, these lasers are said to have a double heterostructure (DH) junction.

DH lasers presently are capable of emitting a hundred milliwatts or more continuously at room temperature. Unfortunately their life is short because of a rapid degradation mechanism; however, the coherent and directional nature of the emitted radiation makes them well suited for several applications.

CHAPTER 2

Infrared LEDs

The first commercial LEDs were made of gallium arsenide (GaAs) and emitted in the near infrared part of the electromagnetic spectrum at about 900 nm. GaAs is still the most-studied light-emitting semiconductor, but recently much of the ongoing research has shifted to GaAs combined with phosphorus or aluminum to give visible red light. This shift in priority is of little concern to GaAs infrared emitters since they are by far the most efficient LEDs available. Even inexpensive silicon-doped GaAs diodes such as the RCA SG1004 may have a power efficiency of as much as 5 percent. More expensive domed diodes may have an efficiency of more than 10 percent.

Technically speaking, it is not correct to describe an infrared emitter as a light emitting diode. This is because "light" is defined as those wavelengths that are visible to the human eye, and infrared is not normally visible (though the eye can see up to one micron when the intensity is high enough). For the purpose of this book, however, light will be defined as visible and *near*-visible radiation to prevent confusion. Sometimes infrared- (IR) emitting diodes are termed IREDs, but in this text both visible and infrared emitters will simply be called LEDs.

Applications for GaAs diodes include voice communications, intrusion alarms, infrared photography, optical radar, control circuits, optoelectronic couplers, and covert illumination. Control circuits alone are made in literally dozens of configurations ranging from automatic end-of-tape detectors for recording machines to broken-beam counting systems. The potential for optical communications is equally as great, with impressive demonstrations of television transmission over a 30-mile link by means of a GaAs diode as early as 1962.

The most useful infrared LEDs are those made from GaAs. While some diodes are made from conventionally doped GaAs, others are made from GaAs doped with silicon (GaAs:Si). These units do not have the extremely fast frequency response of straight GaAs, but they have much higher power efficiencies since they are more transparent to the infrared emitted by the diode.

There are other infrared-emitting diodes as well, and this chapter will describe several of them. But most of the discussion will center around the more commonly used GaAs units.

GALLIUM ARSENIDE LEDs

Texas Instruments placed the first commercial LED on the market in 1962. The diode was the SNX100, a flat emitter made of GaAs with an output of 44 microwatts at 100 mA. Now low-cost GaAs emitters are available that emit a full milliwatt at 100 mA. More expensive domed diodes that emit 35 or more milliwatts at higher current levels are also available.

Potentially, GaAs is an exceptionally efficient semiconductor light source. In fact, several studies have shown properly fabricated GaAs LEDs can have an internal quantum efficiency very near 100 percent. Unfortunately, p-type GaAs is not very transparent to the emitted radiation (910 nm), and diodes must be designed to permit the internally generated infrared to pass through the relatively transparent n-type material. The fact that recombination tends to take place at or near the p side of the junction further complicates matters by reducing the light-emitting properties of the diode.

In an effort to improve the external efficiency of GaAs LEDs, a great deal of effort has been expended in reducing the internal reflection problem discussed in Chapter 1. In 1963, Texas Instruments placed a diode formed into a hemisphere on the market. The diode had about twenty times the efficiency of its predecessor, the SNX100. Unfortunately, domed diodes are very difficult to fabricate and, thus, are quite costly. Therefore, most commercial diodes are in the flat configuration, and clear epoxy is used to reduce the internal reflection problem.

While GaAs LEDs do not have a power conversion efficiency as high as desired, they are very easy to make. Most commercial units utilize a relatively simple zinc-diffusion technique to produce a pn junction in an n-type substrate wafer of GaAs. The diffusion process permits numerous individual diodes to be formed simultaneously together with their contacts by means of the straightforward photolithographic techniques employed in transistor fabrication.

A typical GaAs LED is the GE SSL-4 shown in Fig. 2-1. This diode provides a power output of 0.4 mW at 100-mA forward current. The peak wavelength is 900 nm, and the spectral band width is about 21 nm. The SSL-4 has an exceedingly fast rise time of less than 5-nanoseconds. Recently GE introduced the SSL-54, a high-powered version of the SSL-4. The new diode employs an integral reflector to obtain more light emission. It has a power output of 1.0 mW at 100 mA, more than twice that of the SSL-4.

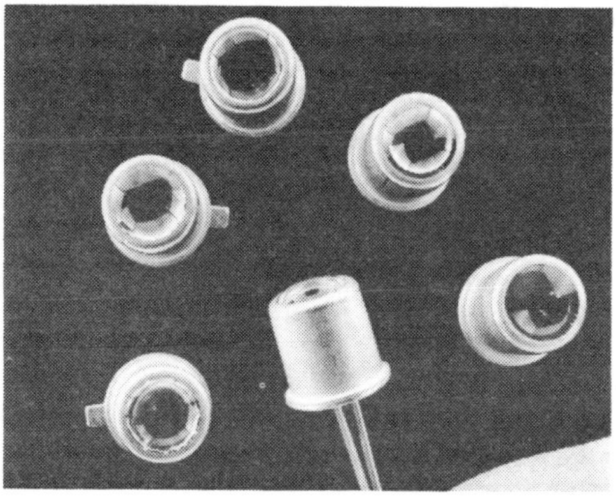

Courtesy General Electric Co.

Fig. 2-1. SSL-4 gallium arsenide infrared emitters.

GALLIUM ARSENIDE PLUS SILICON LEDs

Gallium arsenide can be doped with silicon to give both n- and p-type material, and as such it is called an *amphoteric* dopant (from a Greek word meaning *both*). The process is temperature dependent. As the GaAs drops below 900°C during processing, the silicon atoms occupy gallium sites in the crystal. Below 850°C the silicon atoms occupy arsenic sites. As a result of the temperature effects on material processing, it is possible to form epitaxial layers of n- and then p-type material on a GaAs substrate. Normally, LED junctions are formed by causing an epitaxial layer of p-type GaAs plus silicon to be formed on an n-type substrate of the same material.

The addition of silicon to GaAs causes the emitted light to be at *less* than the bandgap energy. At room temperature the emission will be centered at about 940 nm which corresponds to an energy of 1.32 electron-volts or 0.05 electron-volt below the bandgap energy. Since the infrared has a lower energy, it is not absorbed by the semiconductor to the same extent that 910 nm radiation is absorbed by nonsilicon-doped GaAs. The result is an LED with a much higher external quantum efficiency than straight GaAs. Reduction of internal reflection at the crystal borders by means of special geometries, antireflective coatings, or plastic coverings will provide an additional improvement in efficiency.

While GaAs:Si LEDs are more efficient than their GaAs counterparts, they exhibit much slower response times. A GaAs diode can emit light within a nanosecond of an applied current pulse, while a GaAs:Si unit may require 250 to 300 nanoseconds. Fall times are virtually identical, with GaAs requiring as little as a nanosecond to fall from full to 10 percent radiance and GaAs:Si requiring 200 to 240 ns.

The most common theory to explain the delay mechanism in GaAs:Si diodes has to do with so-called "trapping centers." These centers must be filled with excited electrons before significant external light emission can take place, and the time required to fill them is responsible for the delay. At the end of an applied current pulse, the electrons held in the trapping centers are released to combine with holes and in the process emit light. The traps require a finite time to release excited electrons, and the result is a long fall time. It should be noted that GaAs diode lasers, depending on the configuration, can also exhibit delays due to trapping centers.

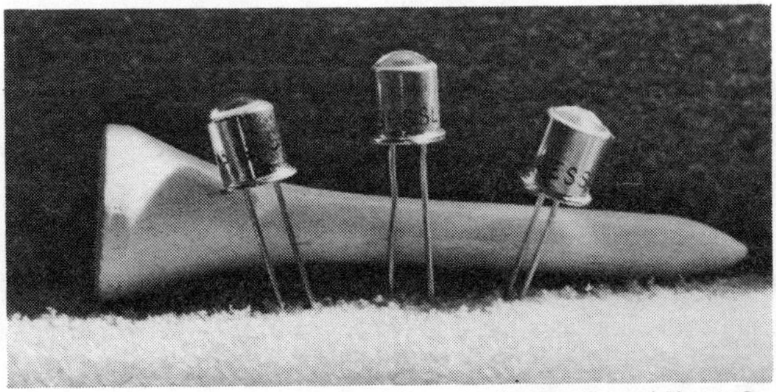

Courtesy General Electric Co.

Fig. 2-2. SSL-5 gallium arsenide silicon infrared emitters.

As a result of its slow rise and fall times, GaAs:Si has a frequency response of only about 100 kHz. This is more than adequate for many applications but nowhere near the 50-MHz capability of GaAs diodes. Obviously, for very high frequency applications GaAs is the better of the two devices. But for the many applications where slower frequency response can be tolerated, the more efficient GaAs:Si is superior.

A typical GaAs:Si LED is shown in Fig. 2-2. This diode, the GE SSL-5C, emits 2.9 mW at 100 mA. Rise time is a slow 300 nanoseconds, but the high output power of the diode makes it suitable for applications where high frequency response is unimportant. GE now offers an improved version of the SSL-5C, the SSL-55C. The new diode employs a smaller chip size and an integral reflector to achieve more than twice the output of the older unit. At 100 mA, the SSL-55C delivers 6 mW.

GALLIUM ARSENIDE vs GALLIUM ARSENIDE PLUS SILICON

Several important considerations apply when choosing an infrared emitter. As was noted earlier, GaAs:Si is considerably more efficient than straight GaAs. Unfortunately, the increase in efficiency is at the cost of frequency response, and that is why a careful consideration between the two materials must be made before choosing an infrared source for a particular application.

In many cases, trading off GaAs and GaAs:Si devices is simple. For example, if a frequency response in excess of a megahertz is required, only GaAs should be considered. And GaAs:Si is a logical choice for high-power, low-frequency-response applications. Problems occur when the infrared source must have both high power-output capability and fast response time. At present the only solution to this dilemma is the use of a high-powered GaAs dome emitter or an array of several GaAs flat diodes.

Other tradeoffs are more subtle. The infrared emitted by GaAs:Si diodes is somewhat more susceptible to absorption by water in the atmosphere than the emission of GaAs. This effect is shown in Fig. 2-3. Even though GaAs:Si LEDs are subject to more atmospheric absorption than GaAs diodes, the high power available from the silicon-doped devices makes them the best choice for long-path-length, medium-frequency applications.

The wavelength of GaAs:Si diodes creates two additional problems. One concerns the spectral width of the material. Since GaAs:Si has a spectral width almost twice that of GaAs, narrow-bandpass optical filters can be more efficient with GaAs. The other wavelength problem concerns the availability of suitable narrow-bandpass filters designed

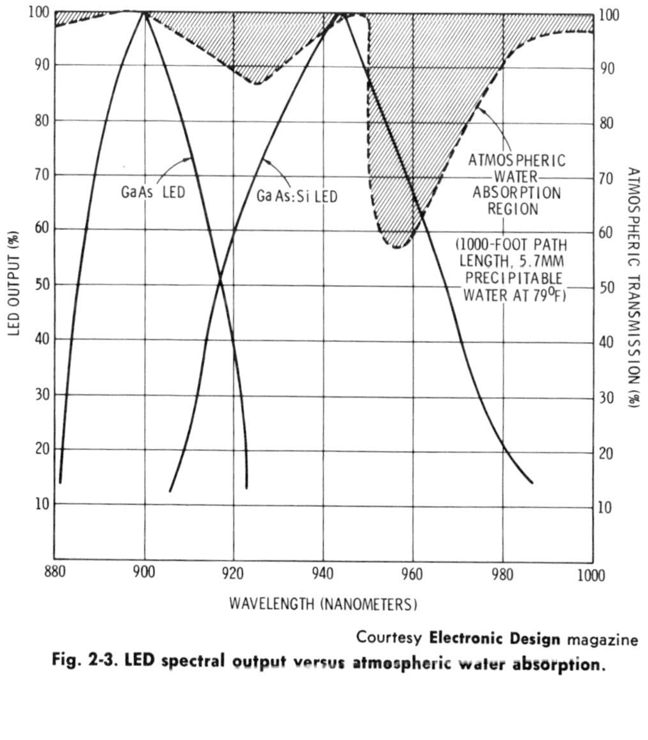

Courtesy **Electronic Design** magazine

Fig. 2-3. LED spectral output versus atmospheric water absorption.

to permit only the wavelength of interest to enter receivers for diode light source communication links and optical detection systems. Since both GaAs diode lasers and noncoherent infrared sources emit at a wavelength of 910 nm, many optical filter manufacturers offer commercial filters at this wavelength. Filters for GaAs:Si LEDs are not as readily available and may even require special ordering. The result can be higher cost and procurement delays.

A good way to evaluate the relative merits and demerits of GaAs and GaAs:Si is to compare the specifications of commercially available units, and a good choice is the GE SSL-54 and the SSL-55C. Both diodes are physically identical and have junction areas of 400 square microinches, but the SSL-54 is straight GaAs while the SSL-55C is silicon-doped. The SSL-54 has a slightly lower forward voltage than the SL-55C (1.25 and 1.35 volts respectively). The SSL-54 also has a lower wavelength (900 and 940 nm). The power output of the two diodes is a more significant feature, and the SSL-55C has a typical output of 6.0 milliwatts at 100 milliamps as compared to only 1.0 milliwatt for the SSL-54. But while the SSL-54 has a lower power-output rating, its frequency response is much faster than that of the SSL-55C. While the SSL-55C has a rise time of 300 ns, the rise time of the SSL-54 is less than 10 ns. These specifications, along with several others, are summarized in Table 2-1.

Table 2-1. GaAs:Si LEDs

Characteristic	SSL-54	SSL-55C
Semiconductor	GaAs	GaAs:Si
Forward voltage	1.25 V	1.35 V
Power output:		
100 mA continuous	1 mW	6 mW
1 A at 1 μsec	10 mW	38 mW
10 A at 1 μsec	75 mW	240 mW
Peak spectral emission	900 nm	940 nm
Rise time	10 ns	300 ns
Fall time	10 ns	200 ns

OTHER INFRARED LEDs

Wavelengths above 1 micron (1000 nm) can be obtained with gallium indium arsenide (GaInAs). Several commercial GaInAs emitters with a peak wavelength of 1.06 microns are available. These LEDs are valuable for use in evaluating detector and receiver systems for neodymium-doped glass and YAG laser systems which also emit at 1.06 microns. Spectronics makes a line of commercial 1.06 micron emitters that have a calibrated output. The devices typically emit 20 microwatts at 50 mA from a source 0.01 inch in diameter.

Several lead compounds are efficient emitters of infrared. Soviet scientists have described a simple technique for obtaining 2.5-micron radiation from ordinary lead sulfide (PbS) infrared-detecting photoresistors. The scientists operated the PbS photoresistors with 100 to 800 volts in pulses 0.4 to 1.0 millisecond wide. The radiation was detected by another PbS detector.

Lead selenide (PbSe), lead telluride (PbTe), and other lead compounds have been used to make LEDs and lasers that emit at wavelengths of up to 30 microns. Some emitters operate at 10.6 microns, the wavelength of carbon dioxide lasers. Lead/tin (PbSn) alloys can be used to produce infrared emitters and lasers with interesting and unusual properties. PbSnSe, for example, can be used to make pn junction lasers with very low thresholds. These lasers emit at 34 microns, but the wavelengths can actually be changed or tuned by pressure, temperature, or even a magnetic field. These lasers and emitters are very useful in infrared spectroscopy.

COMMERCIAL IR LEDs

More than a dozen manufacturers offer commercial infrared emitting diodes. The devices range from tiny coaxial pill packages designed for punched card reading to high-power domed emitters. Several different diode configurations are shown in Fig. 2-4, and there are numerous others.

The tiny size of a hermetically sealed TIL23 is emphasized in Fig. 2-5 by comparison with the head of a pin. These tiny diodes are extremely rugged and are designed for use in computer punched card readouts. Several miniature LEDs are shown on a postage stamp in Fig. 2-6. Two TIL23s are in the center, on the left is a GE SSL-15, and on the right is an RCA SG1004. Diodes as small as these are excellent for many applications, but they sometimes prove difficult to install. The TIL23 is housed in a metal package with a ceramic seal and a glass lens, but the SG1004 is sealed in epoxy resin. While both

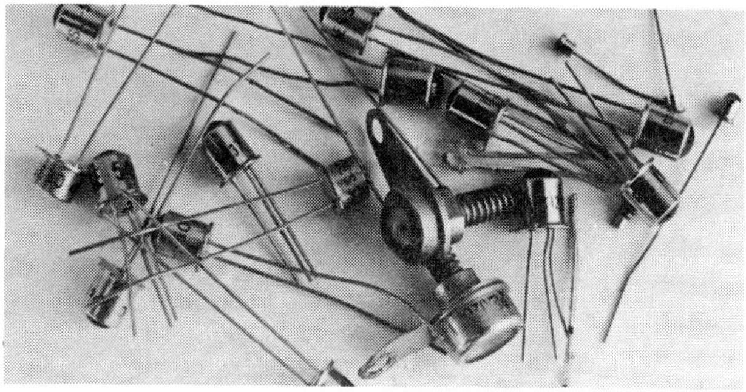

Fig. 2-4. Assortment of commercial infrared LEDs.

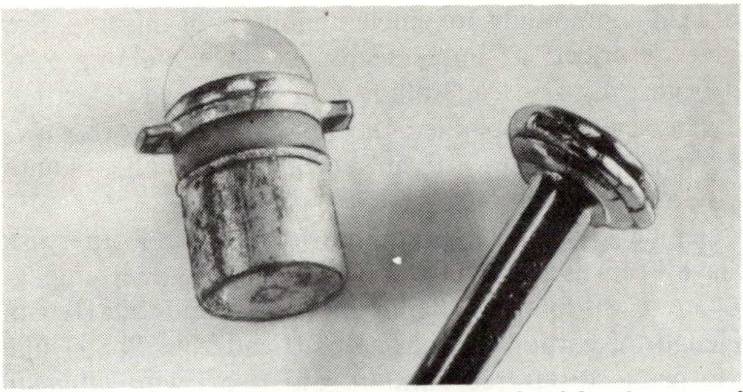

Fig. 2-5. Texas Instruments TIL23 LED shown with pinhead for size comparison.

devices can be dip soldered to a circuit board if temperature ratings are not exceeded, RCA recommends the use of a conductive epoxy for installing its diode.

The internal construction of several small diodes is shown in Fig. 2-7. Notice the integral reflector in the SG1004. Other infrared diodes also employ reflectors and several will be described shortly.

Many common infrared LEDs are housed in TO-18 style transistor cans with a flat glass window or lens in the end of the package. Fig. 2-8 shows a typical TO-18 package with a lens and a modified TO-18 with a flat glass window. The diodes in the figure are, respectively, an

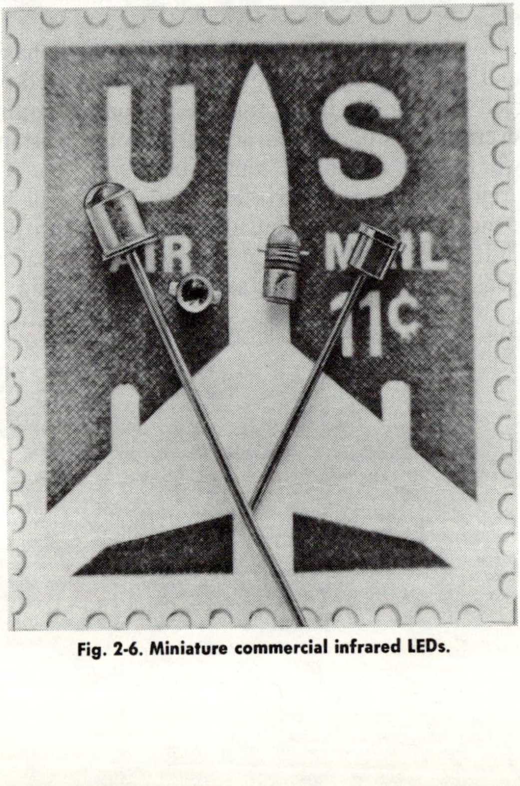

Fig. 2-6. Miniature commercial infrared LEDs.

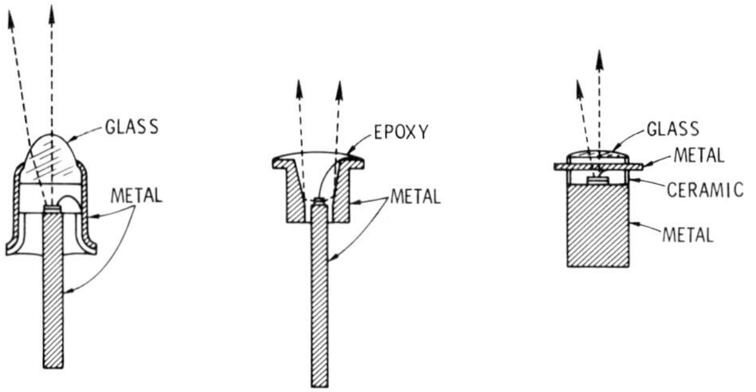

(A) Spectronics SE1450. (B) RCA SG1004. (C) Texas Instruments TIL24.

Fig. 2-7. Construction details of miniature IR LEDs.

SSL-55C and SSL-55CF. Infrared beam patterns from these diodes appear elsewhere in this chapter.

A very popular packaging technique for visible LEDs is the epoxy housing. The epoxy provides a sturdy support for the LED leads, an integral lens, and a good index of refraction match to the LED chip. And epoxy mounting is much faster and cheaper than the metal technique. An enlargement of a typical epoxy infrared emitter is shown in Fig. 2-9. It is inexpensive and has a uniform beam pattern. Continuous forward current is limited to 40 mA because of the lack of heat sinking, and the manufacturer claims a power efficiency of 5 percent.

Texas Instruments exploits the economy of epoxy and the heat sinking of metal with its TIXL27 high-power emitter. Shown in Fig. 2-10, this diode can be operated at a continuous forward current of 300 mA, since it employs a large metal mounting stud. At this current level the diode has a typical power output of 20 mW.

Several firms, including Texas Instruments, Spectronics, and Raytheon, manufacture hemispherical domed GaAs and GaAs:Si high-

Fig. 2-8. Lens and flat-window versions of the General Electric SSL-55C LED.

Fig. 2-9. Close-up photo of Texas Instruments TIL32 encapsulated infrared LED.

Fig. 2-10. Texas Instruments TIXL27 high-power infrared LED.

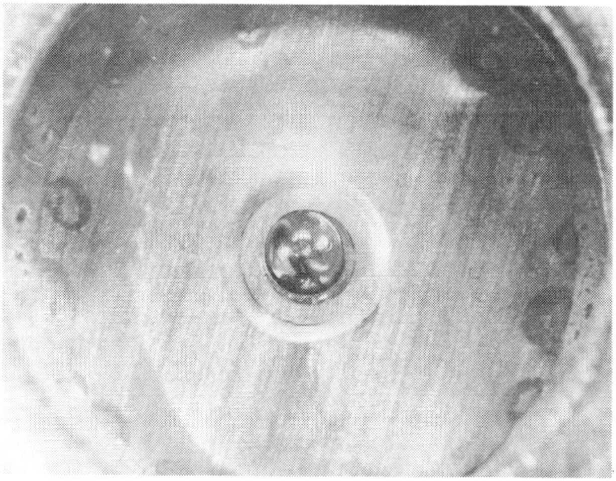

Fig. 2-11. GaAs dome of Texas Instruments high-power infrared LED.

power diodes. With power efficiencies of up to 10 percent and more, these devices are as impressive as they are expensive. Costs range upwards from more than a hundred dollars, with power outputs ranging up to 200 mW or more for continuous operation. Fig. 2-11 is a closeup photograph of the 36-mil-diameter dome of an early Texas Instruments domed emitter, the PEX1201. Table 2-2 summarizes the more important characteristics of some common infrared emitters. The table is by no means complete, since all specifications and manufacturers are not represented. But the table does give a good idea of just what is available.

A word of caution about the table: Because of different techniques of measuring LED parameters and a few attempts at exaggerating power output, the data in the table should be independently checked

Table 2-2. Typical Commercial Infrared Emitters

Type No.	Manufacturer	Power Output	Forward Current
SSL-55C	General Electric	6 mW	100 mA
ME 2	Monsanto	9 mW	1 A
SG1004	RCA	3 mW	50 mA
SE5451	Spectronics	3.5 mW	100 mA
SE6477	Spectronics	11 mW	200 mA
TIL31	Texas Instruments	6 mW	100 mA
TIL32	Texas Instruments	1.2 mW	20 mA
TIXL27	Texas Instruments	20 mW	300 mA
TIXL14	Texas Instruments	60 mW	1A

Note: Data supplied by manufacturers. See company data sheets for additional information on these and newer diodes.

for accuracy. The data comes straight from manufacturer's data sheets. Some simple evaluation techniques will be described later in this chapter.

BUILT-IN REFLECTORS

In order to collect more of the radiation emitted by an LED chip, many diode-mounting packages employ an integral parabolic reflector. The RCA SG1004 diode discussed in the section on commercial devices is an outstanding example of how an integral reflector can enhance power output from an LED. This tiny diode was found to emit more infrared radiation for the same power input than any other nonhemispherical diode in a series of tests conducted by the author. A photographic enlargement of the SG1004 in Fig. 2-12 shows the chip and reflector.

Fig. 2-12. RCA SG1004 high-efficiency LED with built-in reflector.

Another diode that uses an integral reflector is the General Electric SSL-55C. As can be seen in the close-up photograph in Fig. 2-13, the chip is mounted in a circular dimple formed in the base of the LED mounting package and is coated with a drop of clear epoxy resin for more efficient extraction of the radiation. The dimple reflector is so efficient at collecting the edge emission from the LED chip that less GaAs:Si can be used to obtain more power than a diode without the reflector. The SSL-5C, for example, emits 2.9 mW at 100 mA from a chip 25 mils square, while the SSL-55C emits 6 mW at 100 mA from a chip only 20 mils square. Since it requires less semiconductor, the SSL-55C is even cheaper than the lower-power, nonreflector SSL-5C.

Another LED with an integral reflector is the Spectronics SE2430. This diode gives a uniform emission pattern with a half-power divergence of about 30°. Fig. 2-14 shows an array of more than a thousand SE2430s made by Spectronics for use as a wide-area infrared illumination source. A close-up of the array is provided in Fig. 2-15.

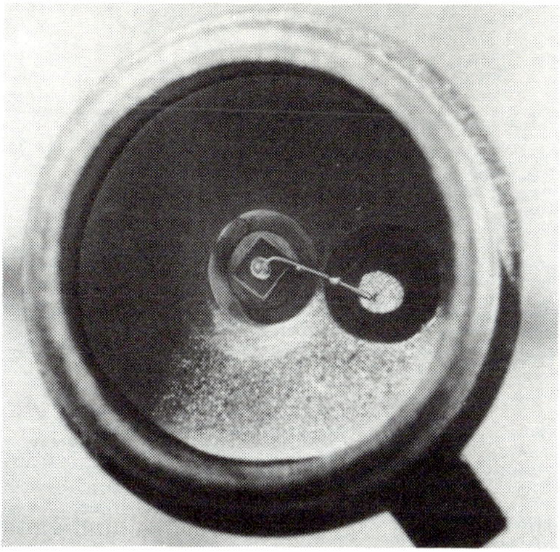

Fig. 2-13. General Electric SSL-55C infrared LED showing GaAs:Si chip (20 mils square) and dimple reflector.

There are other LEDs that employ integral reflectors also. The PEX1201 domed emitter shown in Fig. 2-11, for example, employs a dimple reflector to collect much of the radiation that might otherwise be lost inside the mounting can. Domed diodes are also available with larger, external reflectors. Texas Instruments offers both single reflectors and arrays for use with its range of high-powered emitters.

Fig. 2-14. Spectronics infrared LED array.

Fig. 2-16 summarizes several different integral reflector LEDs. Some are more efficient optically than others, but all improve on the performance of the device without a reflector. Beam patterns of reflector LEDs are generally far more uniform than those from diodes that employ a focusing lens. A combination of lens and reflector such as

Fig. 2-15. Close-up of diodes in Spectronics LED array.

that found in the GE SSL-55C will produce a nonhomogenous beam with a high-power-density, uniform, central spot.

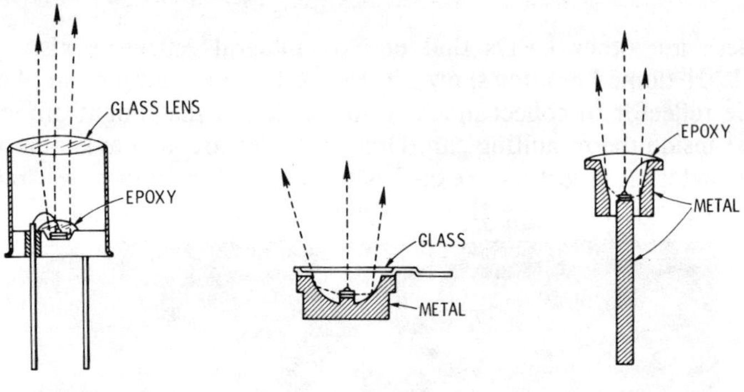

(A) GE SSL-55C. (B) Spectronics SE2430. (C) RCA SG1004.

Fig. 2-16. LEDS with integral reflectors.

LED PULSER MODULES

For many applications, visible and infrared-emitting diodes are simply operated continuously. But by using brief pulses of high current it is possible to obtain far more output power from the LED.

Since relatively simple circuits can be used to current-pulse an LED or even a diode laser, several companies have considered the idea of using integrated circuitry to assemble a complete LED and pulser system on a single tiny substrate.

MERET, Inc., makes integrated LED and diode laser modules. One LED module, the FLP125, operates from a power-supply voltage of from 12 to 50 volts (depending on the power output desired) and achieves peak powers of up to one watt in a 90-nanosecond pulse. The FIP 327 is an injection laser module that emits more than 10 watts in a 40-nanosecond pulse.

As shown in Fig. 2-17, the laser module is very compact. The major drawback to the units is cost, with current prices in the $100 range. Large-scale production of the modules will almost certainly reduce their price.

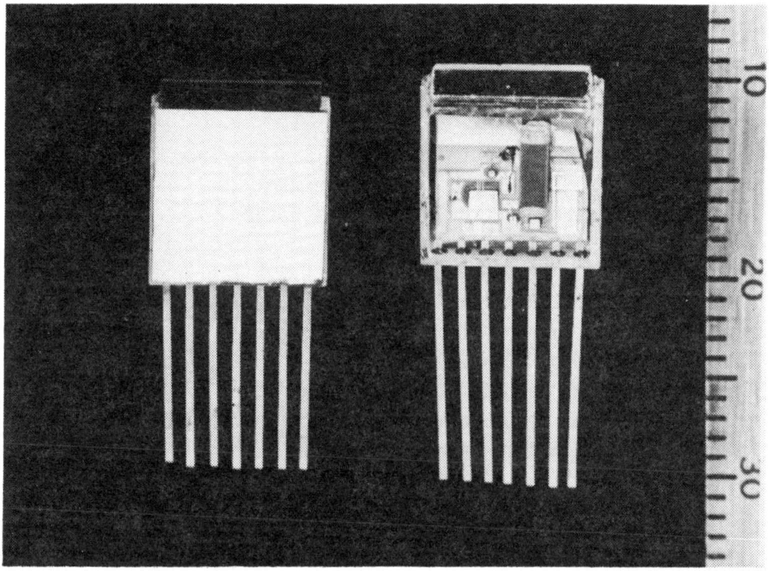

Courtesy MERET, Inc.

Fig. 2-17. MERET, Inc. FIP 327 injection laser/pulser module.

EVALUATING COMMERCIAL IR LEDs

Some manufacturers supply excellent specification sheets for their line of LEDs; others do not. The result can be confusion and problems for the potential customer. In fact, when a large purchase of IR LEDs is contemplated, the purchaser would be well advised to obtain representative product samples from several manufacturers before making a final choice.

Infrared photographs provide an inexpensive, permanent recording of LED beam patterns. Unfortunately, they are not as good for comparing diodes made by different manufacturers as one might think, since one diode might have a different power-output rating than another. As a result, when both diodes are photographed at the same current level, one will appear to be brighter on the film. The pattern from the stronger diode may blur out beam inhomogeneities, or very weak off-axis patterns may be revealed. Before comparing different

LEDs on infrared film, it is a good idea to operate both at the same total output power.

It is easy to operate different LEDs at the same output power with a simple test made with any uncalibrated solar cell. Just place the lens of one of the LEDs directly against the solar cell and set its forward current to about 10 mA. Record the current from the solar cell with a 0–1-mA meter. Next, set the current for the second LED to give the same solar-cell current as the first. To give equal exposure intensity on the infrared film, simply operate both LEDs at the forward-current levels that give an identical solar-cell current. Of course, exposure time for both diodes must be identical or the output matching technique is of no value.

There are several ways to gain a quick impression of LED beam uniformity. One is to use an infrared-sensitive phosphor screen, and this technique is described more fully in Chapter 5. Another is to employ a ground-glass screen and an image converter that changes infrared to visible light. The diode is aimed at the screen, and the converter is used to observe the pattern formed on the ground side of the glass. Image converters are also described in Chapter 5.

More precise knowledge about the beam pattern of an LED requires a profile scan with an infrared detector. Coupled with IR photos of beam patterns, such scans can provide very useful data about LEDs. There are several simple techniques for performing such a scan, and one is described in the following section.

MEASURING LED BEAM PATTERNS

A simple and reliable technique for measuring LED beam patterns is to scan the emission with a linear response infrared detector. A good low-cost choice is a silicon solar cell, but for more sensitive applications a PIN photodiode can be employed. Chapter 5 describes several simple circuits for use with solar cells and PIN diodes that are ideal for beam scanning.

Prepare to make a scan of beam pattern by mounting the LED at the nadir point on a polar graph. The graph can be purchased commercially or it can be made with the help of a protractor and pencil. It is important that the LED be aligned down the 0° axis of the graph (see Fig. 2-18).

The actual scan is made by placing the detector a fixed distance from the LED at the + or − 180° point and recording the measurement of detector output at intervals of several degrees. If a solar-cell detector is employed, its resolution should be increased by blocking all but a narrow 1- or 2-mm stripe of its active surface with black electrical tape. The cell is attached directly to a voltmeter, and readings are recorded as the cell is moved across the graph with the exposed

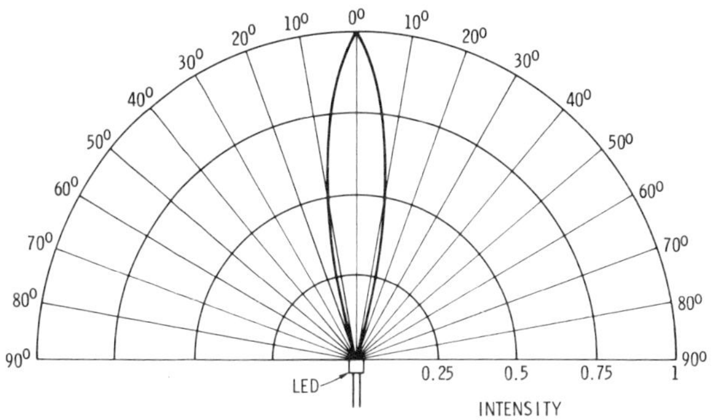

Fig. 2-18. Polar graph of LED power distribution.

stripe *perpendicular* to the paper. Make sure the room lights are either off or subdued, so that their radiation does not cause false signal indications on the meter.

When the scan is completed, the voltage readings are plotted on a sheet of polar graph paper and connected by a pencil line. The result is shown in Fig. 2-18. It may be desired to normalize the graph so that the region of peak intensity near the zero axis is shown as 1.0 (100 percent). This can be done by multiplying each measurement by a correction factor before plotting it on the graph or by moving the detector along the central axis until a meter reading of, for example, 0.1 volts (or 100 percent) is obtained. If the LED has low radiant power and if the meter does not have good resolution at the 0.1-volt level and below, the latter approach may not be suitable. To normalize readings so that peak intensity corresponds to 100 percent, take the highest reading and make a proportion. For example, if the highest reading is 0.20 volt, divide this number into 0.1 to obtain the correction factor of 5. Simply multiply each voltage reading by 5 to obtain measurements normalized to 100 percent.

This technique of beam profiling is simple but effective. There are many ways the technique can be improved; one is to construct a permanent apparatus containing a detector on a movable arm, electrical connections for LEDs to be tested, and an output meter. The experimenter or engineer will no doubt find other ways to improve the measurement technique.

LED POWER MEASUREMENT

The measurement of LED total radiant flux can be a controversial area. This is in large part due to variations in detector calibration accuracy and the measurement techniques themselves. As with visible emitters, the most accurate detectors are calibrated with *standard lamps* having a spectral output and radiant flux traceable to measurements made under carefully controlled conditions by the National Bureau of Standards. Unfortunately, two engineers with access to one standard lamp will most likely arrive at slightly different calibration values for the same detector. The differences may range from a few percent to more than ten percent. These differences are unavoidable because of subtle variations in the output of the standard lamp, temperature changes, and the accuracy of measuring instruments.

Because they are inexpensive, readily available, and have a linear response to incident radiation, solar cells are frequently employed as detectors. Several detector manufacturers offer standard PIN diode detectors that have been calibrated with a standard lamp or other calibration technique. The price of these detectors is about double the price of an uncalibrated unit. Two such firms are United Detector Technology (UDT) and EG&G; each sells a calibrated detector for just under $100. Chapter 5 provides details on how solar cells and other detectors can be calibrated. Even when a well-calibrated detector is available, measurement techniques can cause significant errors in power measurements. Some diodes emit wide-angle beams, and small reflectors placed over the detector must be used to collect all the emitted radiation. Even then, some of the radiation is absorbed by the reflector and some may never reach the detector. Another cause of error is far more subtle and concerns radiation from the LED which has reflected from the active surface of the detector. A silicon solar cell, for example, has a diffuse reflectance of about 4 percent at the wavelength of GaAs and GaAs:Si LEDs. If the LED is placed very close to the detector during measurement—and it usually is—some of the reflected radiation will enter the LED again and be reflected back to the solar cell a second time. If this happens, the cell will give a power reading slightly higher than the actual radiated power of the LED.

RADIOMETRY

Radiometry is the science of measuring radiant energy intensity. When the radiant energy is visible light, the cumbersome but necessary photometric system is usually employed to express radiant intensity. The radiometric system is employed for both visible light and infrared, and it is a far simpler method of measurement.

In radiometry, total optical flux is expressed in watts. When a beam of light containing a total flux of one watt illuminates a spot with an area of one square centimeter, the spot has an incident flux density or *irradiance* of 1 W/cm^2. If the surface of the spot is a perfect reflector, it then has a *radiant emittance* of 1 W/cm^2. The *radiant intensity* of a beam of light is often expressed in watts per steradian. A *steradian* (sr) is the solid angle that defines the surface area of a sphere equal to the square of the sphere's radius (a sphere contains 4π or 12.57 steradians).

Using the radiometric system to explain or understand infrared LED output is straightforward. While the visible emitters must be defined in terms of brightness, intensity, and other terms that are a function of their visibility to the human eye, any infrared emitter can be rated simply by its power output in watts and its beam divergence. Of course, visible LEDs can also be rated radiometrically; as we shall see in the next chapter, 680 lumens of green light at 555 nm corresponds to one radiometric watt. But visible LEDs are almost always used in applications where the human eye is the detector, and therefore their power rating is of much less significance than their visual rating. This discussion of radiometry will be continued in the next chapter when photometry and radiometry are compared.

To understand the use of the radiometric system in rating an infrared LED, consider a diode emitting 2.5 mW into a one-steradian beam. The diode is said to have a radiant intensity of 2.5 mW/sr. Most manufacturers specify LEDs strictly in terms of power output and whatever divergence the diode may have; however, in many applications it is more important to know the radiant intensity of the beam, unless external optics will be used to alter the beam. One manufacturer that rates its infrared LEDs in terms of radiant intensity is Fairchild. The specification sheet for the Fairchild FLD100 lists the total power output (1.6 mW at 50 mA) and the radiant intensity (0.4 mW/sr at 50 mA).

Radiant intensity is particularly important when a detector is to be used to sense the radiated power of an LED. If the LED is specified in terms of power density, then irradiance at a detector can be easily calculated by means of the expression

$$H_E = \frac{P/sr}{d} \qquad (2\text{-}1)$$

where,

H_E is the irradiance at the detector surface in mW/cm^2,
P/sr is the radiant intensity of the LED in mW/sr,
d is the distance in cm from the LED to the detector.

Since the radiant intensity of the FLD100 is known, it is a simple matter to calculate the irradiance at a detector 10 cm away. Inserting the appropriate values into Equation 2-1, we find that the detector is irradiated with a power density of 0.04 mW/cm^2. If this is not sufficient radiant power to properly operate the detector, the LED can be pulsed to give higher peak power, or a lens can be used to focus more of the available power from the LED onto the active surface of the detector. As will be shown in Chapter 6, the lens can be placed at either the LED, the detector, or both.

CURRENT-PULSING LEDs

The total flux available from an LED is directly related to the forward-current density. With most LEDs, power output increases linearly with current until a saturation point is reached. At this point the temperature of the diode has risen to a degree where further power output becomes progressively less efficient. If the current level is raised too far beyond this point, the diode may be destroyed.

It is possible to increase the current level—and hence the power output—of an LED beyond its rated value by employing an external heat sink. So long as the heat sink keeps the diode at or near room temperature, the diode current can be increased to a certain value above the normal rating. Sometimes a coolant such as dry ice or liquid nitrogen is employed to remove heat from the diode.

A very simple way to obtain ten or more times the rated dc power output of an LED is to apply very brief pulses of high current. If the pulses are of an appropriate width and separation to give a *duty cycle* (diode "on" time) within the manufacturers ratings, the diode will not become warm enough to cause catastrophic damage. To see how effective current pulsing an LED can be, consider Table 2-3. The table compares dc and pulse operation for an RCA SG1004 infrared LED.

Table 2-3. RCA SG1004 LED

Forward Current	Power Output
15 mA continuous	1 mW
50 mA continuous	3 mW
1.0 A, 2 μsec	40 mW
1.4 A, 2 μsec	49 mW

When operated at dc, the maximum permissible current level is 50 mA with a typical power output of 3 mW. However, when driven with 1-A pulses 2-microseconds wide, the same diode will emit 40 mW. This is an increase of more than thirteen times the value obtained with operation at 50 mA.

At 1.5 amperes, the maximum allowable duty cycle of the SG1004 is 2 percent. The duty cycle can be increased by reducing the forward current of the diode. Duty cycle can be calculated by multiplying pulse width by repetition rate (frequency). For example, an LED pulsed with 10-microsecond current pulses at a rate of 100 Hz has a duty cycle of 1×10^{-3} (0.001) or 0.1 percent. The *average* power of the diode will then be 0.1 percent of the peak power per pulse.

HIGH-CURRENT PULSER CIRCUIT

A simple way to obtain a high-current pulse is to discharge a capacitor through a low impedance. This technique requires a switching device to connect the capacitor to the LED, as shown in Fig. 2-19. In operation, the mechanical switch is normally off and the capacitor charges through the resistor. When the capacitor has become charged, the switched is depressed and the capacitor voltage is applied directly across the LED.

While the circuit of Fig. 2-19 is elementary and not necessarily practical, it does provide a simple way to obtain some important knowledge about LED pulse characteristics, First, we know that the time constant of a charging circuit is expressed as

$$T = RC \tag{2-2}$$

where,

T is time constant in seconds,
R is resistance in ohms,
C is capacitance in farads.

If the basic circuit has an R of 10,000 ohms and a C of 0.1 microfarad, then its charging time constant is 0.001 second. In other words, the capacitor will require 1 millisecond to charge to 63.2 percent of its final value.

Turning this around, we can measure the time it takes for the capacitor to discharge through the LED and determine the *dynamic* resistance of the LED. Dynamic resistance may be different than the "off" or *static* resistance of the device. If an oscilloscope connected across the LED reveals a pulse about 150 nanoseconds in duration, then solving for R gives a dynamic resistance of about 0.6 ohm. Since the dynamic resistance of the LED is known, it is possible to calculate peak current through the LED. From Ohm's law we know that current equals voltage divided by resistance. If the capacitor is charged to a

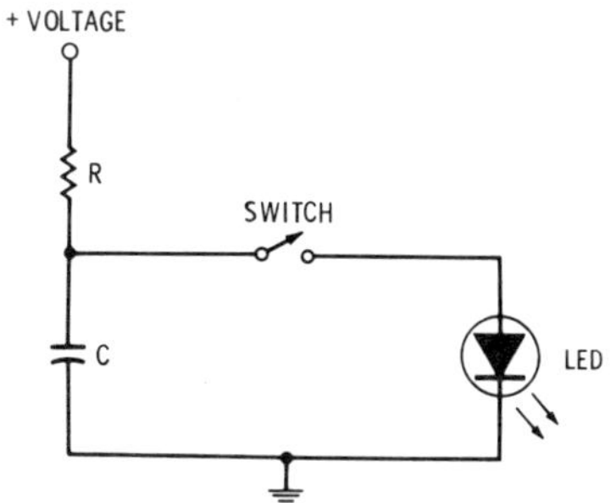

Fig. 2-19. Basic pulse generator.

voltage of 12 volts, then the peak current through the LED (assuming no circuit losses) would be 2 amperes. In practice, peak current will generally vary somewhat from this value. When peak current through an LED must be known in order to prevent exceeding the ratings of the device, it is a good idea to connect a small value resistor in series with the LED. By monitoring the voltage across the resistor during a current pulse, it is possible to calculate the current through the resistor and hence through the LED. For example, if our hypothetical circuit incorporated a series resistor of 0.1 ohm and the voltage across the resistor was 0.18 volt, then peak current is 1.8 amperes.

The basic circuit of Fig. 2-19 can be converted into a practical circuit by replacing the mechanical switch with a fast electronic switch such as a transistor, SCR, four-layer diode, or thyratron. If the device shows negative-resistance characteristics, the electronic switch can be made to operate as a relaxation oscillator. Fig. 2-20 shows one possible circuit where a four-layer diode is inserted into the basic circuit and the supply voltage is increased somewhat. In operation the capacitor charges until the breakdown voltage of the four-layer diode is reached. At this point the diode switches on (thus assuming a negative-resistance state), and the capacitor discharges through the LED in a brief, high-current pulse.

Four-layer diode pulse circuits can be made very small. Their chief advantages are simplicity and high repetition rate, but their very low "on" resistance means pulse width is very narrow unless a capacitor of a few microfarads is employed.

Four-layer diodes can cost as much as several dollars or more and an economical way to construct a very simple and inexpensive high-current pulse generator is to employ a silicon switching transistor operated in its avalanche mode. Many common and inexpensive npn transistors are well suited for this purpose, and representative units in-

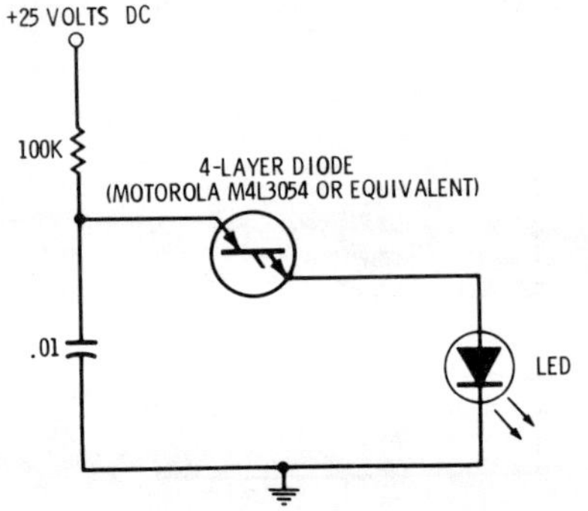

Fig. 2-20. Basic four-layer diode LED pulser.

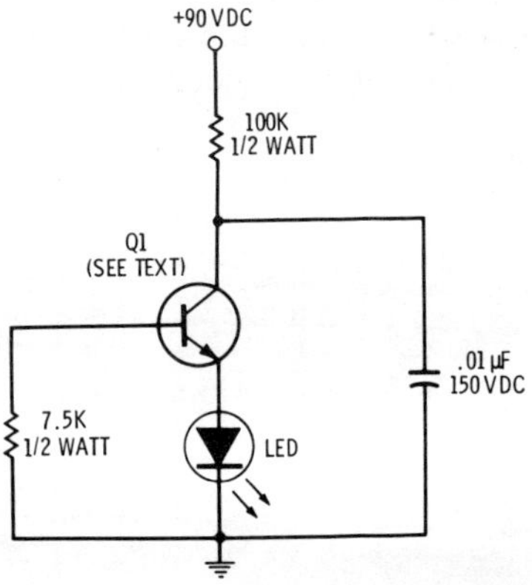

Fig. 2-21. Avalanche transistor LED pulser.

clude the following: 2N914, 2N2222, 2N3643, 2N4400, and 2N5188. Fig. 2-21 shows a circuit for the avalanche-transistor pulse generator. Operation of the circuit is similar to that of the four-layer diode version. The capacitor charges until the collector-emitter breakdown voltage (BV_{cex}) of the transistor is reached and then discharges through the transistor and the LED. The 7500-ohm resistor from the base to ground provides a small base current. As with the four-layer diode unit, very fast, high-current pulses can be generated. The main advantage of the pulser is its simplicity and very low cost. Without a power supply, an entire pulser, including a GaAsP red LED, can be assembled for less than $1.50.

CHAPTER 3

Visible LEDs

A simple but beautiful experiment is to observe a yellow LED through a 50-power microscope while slowly increasing the diode current. With no current, the LED chip, when viewed from the side, resembles two slabs of yellow and orange gelatin. The yellow layer is the n side, the orange the p side, and the border between the two is the junction of the diode. Both layers are transparent.

At a few milliamperes of forward bias, the yellow side of the junction begins to emit a soft yellow glow. As the current is slowly increased, the entire n side of the LED emits a yellow light. Finally, at maximum current of about 50 mA the yellow penetrates into the absorbing p side as well, and the entire diode emits a bright yellow light.

Numerous semiconductors emit visible light when stimulated with light, heat, electricity, or other external forces. The development of the GaAs semiconductor laser in 1962 stimulated a search by scientists for other semiconductors that would support laser action in a junction diode. A major goal was the identification of materials that would give laser action at visible wavelengths at room temperature. While this elusive goal has not yet been attained, the search for materials has led to more than a dozen semiconductors well suited for visible LEDs. This chapter will present many of the best-known materials under subheadings denoting the chief colors: red, orange, yellow, green, and blue. Additionally, there will be discussions on multiple-color LEDs, visual selection and evaluation of commercial LEDs, manufacturing techniques, and commercial devices. Before all this, however, let us digress for a few pages and discuss *photometry,* the science of measuring visible light. As will soon be obvious, an introduction to photometry is quite valuable in accurately evaluating the optical properties of visible LEDs.

LEDs AND PHOTOMETRY

When matching a photodetector to a particular LED, the designer generally gives first consideration to the spectral response of the detector. GaAs infrared LEDs, for example, match the peak spectral response of most silicon detectors quite nicely.

In the case of visible LEDs, the detector is most frequently the human eye, and its spectral response is the determining factor. The retina of the eye incorporates two kinds of light-sensitive cells: rods and cones. Rods are scattered throughout the retina and are more sensitive than cones. Their sensitivity permits night vision, but they are not color sensitive. There are fewer cones than rods, and most of them are localized in a small depression near the center of the retina, the *fovea.* The high concentration of cones in the fovea provides excellent resolution. Furthermore, the cones provide us with the ability to see colors.

Little is known about the complex processes in which the brain transforms signals from the retina into conscious images in full color. However, it is possible to measure the color response of an individual's eyes and plot the results on a graph. The measurement of eye spectral response and light spectral output is what the science of photometry is all about.

The human eye responds to a very small part of the electromagnetic spectrum extending from about 400 to 720 nm. This region is called visible light, and Fig. 3-1 shows the relative position of light with respect to ultraviolet, infrared, radio waves, and other components of the electromagnetic spectrum.

Through processes that are not completely understood, the eye and and brain have the ability to convert light into colors of various hues ranging from violet to red. Fig. 3-2 shows the relation of wavelength to hue as perceived by the human observer. The range of "visible light" is not necessarily restricted to the 400- to 720-nm range commonly given as the limit of human vision. Many observers have noted that the supposedly invisible 910-nm infrared radiation from GaAs diode lasers is visible as a dark red light when the laser is viewed directly against a dark background. The laser light is visible because of its very high intensity. Therefore, for most practical purposes the limit

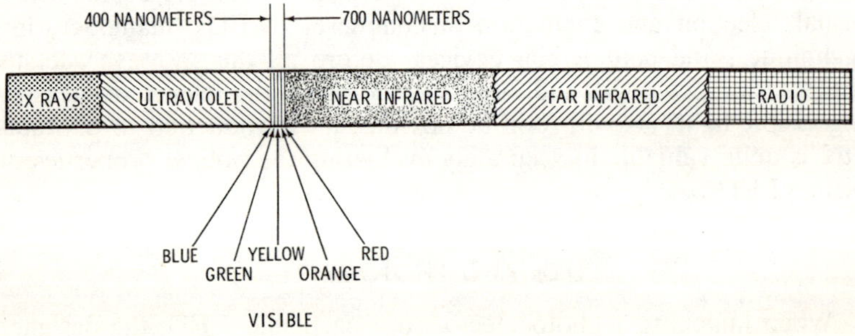

Fig. 3-1. The electromagnetic spectrum.

of visual response is the 400- to 720-nm spread given by most references.

The heart of the photometric system as applied to visible LEDs is the *photopic luminosity* curve standardized by the *Commission Internationale de l'Eclarage* (C.I.E.). This graph shows the relative sensitivity of the nondark-adapted human eye (cone vision) to the visible spectrum and is the key to relating LED color output to the sometimes confusing photometric system.

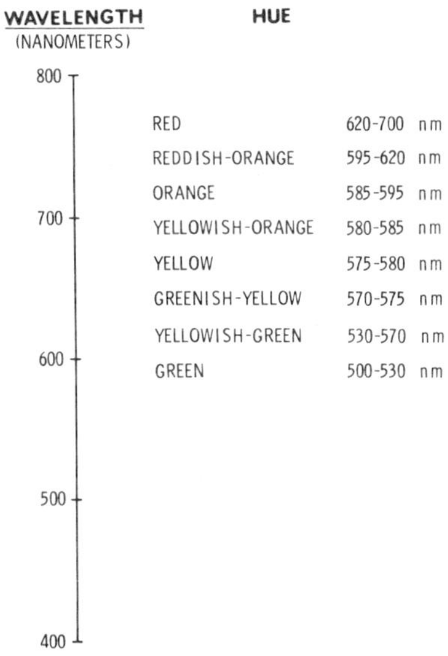

Fig. 3-2. Wavelength versus hue for available visible LED spectral range.

The C.I.E. has another curve for human eye spectral response and it is shifted slightly upward from the photopic curve. This second graph of eye spectral response, the *scotopic luminosity curve,* defines the visual limits of the dark-adapted eye. Since the dark-adapted eye enhances rod rather than cone vision, color response is low, and the scotopic curve is therefore not applicable to LEDs.

According to the C.I.E. photopic luminosity curve, which is shown in Fig. 3-3, the "standard observer" has a peak visual response to light with a wavelength of 555 nm. The eye response is down by 50 percent at 510 nm and 610 nm, the half-power points on the curve. Since the "standard observer" is in reality a nonexisting composite eye made up from an average of about 200 actual test subjects, some individuals may exhibit a slightly altered spectral response. The differences are usually minor, however, and the standard observer's curve is generally a good rule to follow.

Because more and more LEDs are finding use in consumer applications, there has been more interest in people whose visual response is somewhat different from that of the standard observer. Recently Bell Telephone Laboratories found that one of every 50 males experiences reduced sensitivity to red light produced by the most common visible LEDs. In the case of gallium phosphide red LEDs, sensitivity is 14 percent of the normal photopic response, and sensitivity is only 10 percent of normal for gallium arsenide red phosphide emitters. The difference in response is due to the wavelength difference of each

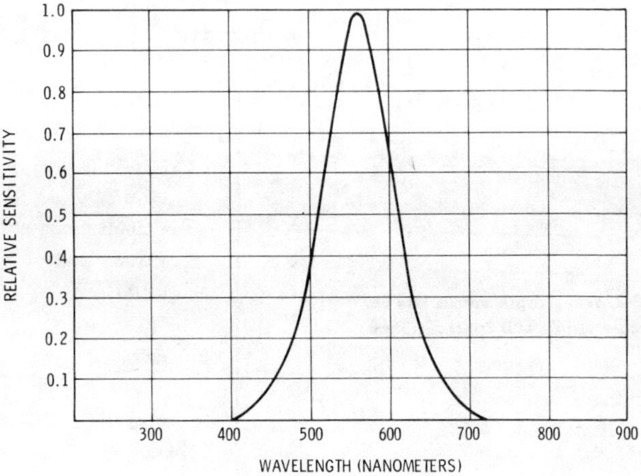

Fig. 3-3. Photopic luminosity curve.

diode. Bell found that there was little or no reduction in response to green LEDs by the subjects with reduced red sensitivity.

Since LEDs are finding such wide use, the Bell study indicates that efficient green emitters would be a valuable development. Besides being more visible to all of the population, they would be easily seen by those males with reduced red sensitivity.

PHOTOMETRIC TERMINOLOGY AND DEFINITIONS

If a visible LED is used to activate an electronic detector, a knowledge of its size and radiometric power output would be adequate for many applications. However, when the LED is to be observed by the human eye, other factors come into play and a variety of photometric terms are used to express them. Some of the more important photometric terms follow.

Total Flux

The fundamental unit of optical flux in photometry is the lumen. At 555 nm, the peak of the photopic luminosity curve, 680 lumens corresponds to one watt. Knowing this, we can superimpose the pho-

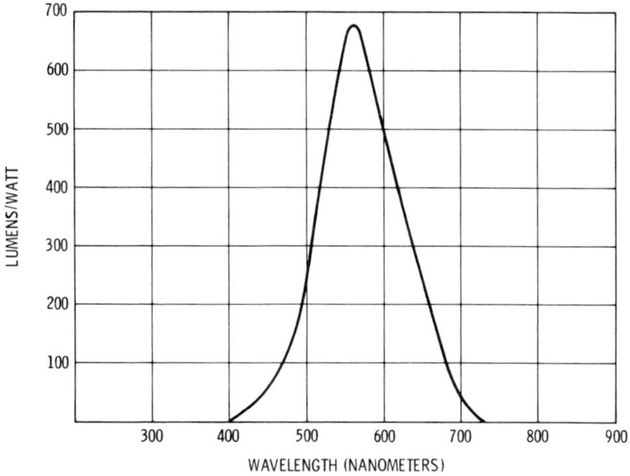

Fig. 3-4. Photopic luminosity curve (lumens/watt).

topic luminosity curve onto a graph showing the relative values of radiant and luminous flux, as in Fig. 3-4. This conversion factor permits comparison of the relative power efficiency of an assortment of LEDs with various color outputs. It would be more convenient if one lumen at 555 nm equaled one radiometric watt, but the photometric system evolved from a measurement technique wherein a sperm candle was the basic standard. The light emitted by one such candle was defined as 4π lumens.

Incident Flux Density

Incident flux density expresses the amount of radiation per unit area illuminating or irradiating a surface. In photometry it is expressed in terms of lumens/cm^2, while watts/cm^2 are used in radiometry. Incident flux density is frequently used in expressing the amount of flux received by a detector. Knowing the minimum sensitivity level of the detector, a knowledge of incident flux density will indicate whether or not a detector will detect an incoming signal.

Emitted Flux Density

In radiometry, emitted flux density is expressed as watts/cm^2, while in photometry it is lumens/cm^2. Of course, other area terms such as square feet or square meters can be employed. Emitted flux density is usually used to describe light reflected from a surface. The reflectance of the surface, which usually varies with the wavelength of the incident radiation, determines the total radiant or luminous emittance. Therefore, a luminous emittance of 4.2 lumens/cm^2 will result when a surface with a reflectance of 50 percent is illuminated with a beam containing 8.4 lumens/cm^2 at the surface. Fig. 3-5 illustrates the definitions of incident and emitted flux density.

Source Intensity

Source intensity is important in both radiometry and photometry since it defines the flux density which will appear at a distant surface. Source intensity is expressed in terms of the flux contained in a solid angle projected in front of a source. In photometry and radiometry this becomes lumens/steradian and watts/steradian, respectively.

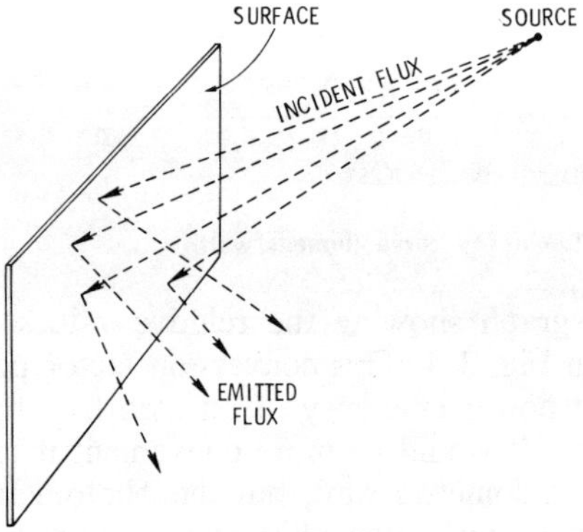

Fig. 3-5. Incident and emitted flux.

The measure of luminous intensity is the candle or *candela.* A light source emitting one candela in all directions, an *isotropic radiator,* emits a total flux of 4 π lumens.

A more specific measure of source intensity can be defined by specifying the luminous (photopic) or radiant (radiometric) intensity per solid angle as a function of area. For radiometry this becomes watts/cm^2/sr (radiance) and for photometry lumens/cm^2/sr (luminance). Luminance is so important in rating and evaluating visible LEDs that it will be described in more detail.

Luminance

The luminance of a visible source is its photometric brightness. Luminance is obtained by dividing the luminous intensity at a given point by the *projected area* of the source at the same point.

Fig. 3-6 shows the definition of luminance or brightness in diagram form. Note that luminance is greatest when a source is viewed straight on, because that is the angle that gives the largest projected surface area of the source. Of course, this condition is modified if the source has a surface that is not flat or that emits in an unusual, nonuniform pattern.

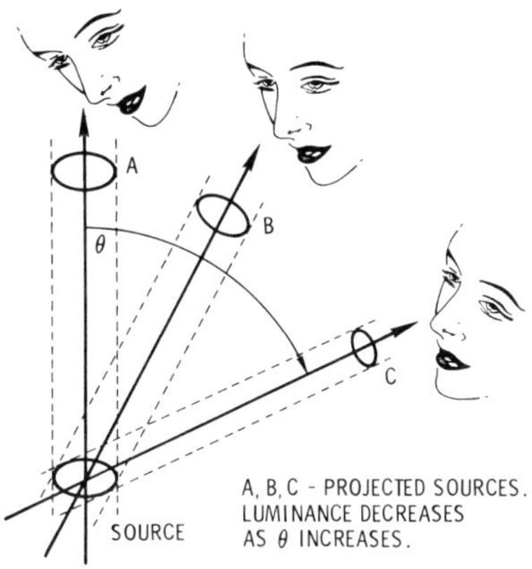

Fig. 3-6. Definition of luminance.

Luminance Terminology

A variety of strange terms have evolved to define various conditions of luminance. In the United States the most common term is the *foot-lambert* (fL). One footlambert corresponds to a luminance of $1/\pi$ candela/ft^2. The *candela,* remember, is the unit of luminous intensity.

While luminance is equated with photometric brightness, it can be misleading to describe luminance as a figure of merit for brightness. This is because the human eye is a very subjective detection system and subtle differences in two light sources with the same luminance may cause one to appear brighter than the other. Physically identical LEDs can safely be compared by equating their luminance with brightness, but different LEDs must be compared either subjectively (direct viewing) or in terms of luminance. This subject will be continued later in this chapter in a section on evaluating visible LEDs.

Summing Up

These basic photometric definitions are of great importance in evaluating and rating visible LEDs. There are other illumination conditions as well, and photometric definitions have been created to define them.

RED LEDs

The first practical LEDs emitted red light centered at about 660 nm and were made from gallium arsenide phosphide (GaAsP). This material is more developed for commercial devices than any other. Typical commercial devices have brightness values of 1000 fL or greater at a drive current density of only about 10 amperes/cm^2. In practical devices, brightness can range from 200 fL at 3 mA for Monsanto's MV-55 to 5000 fL at 1 ampere for the firm's MV-4. Of course, the

MV-55 has a much smaller chip size than the MV-4. Internal quantum efficiency can approach 100 percent, but absorption by the semiconductor and internal reflection cause a significant reduction in external quantum and power efficiency.

GaAsP LEDs are available from at least thirty companies in dozens of configurations. Some tiny units are entirely encapsulated in red or clear epoxy and cost less than a quarter each in large quantities. Others are assembled in arrays and used as alphanumeric readout devices.

While GaAsP LEDs are the units most commonly available, gallium phosphide (GaP) units are several times more efficient on a photon per electron (quantum efficiency) basis. The 690-nm wavelength of GaP falls even lower on the photopic luminosity curve than that of GaAsP with a resultant luminous efficiency of only 3 percent (vs 6 percent for GaAsP). But the overall brightness (fL) vs current density is about 2.4 times that of an equal size GaAsP diode.

GaP has several important disadvantages. From an operating standpoint the most significant disadvantage is that the light output does not continue to increase with applied current as with most other LED materials. At relatively low current densities the material becomes saturated, and the light output ceases to rise. Depending on the doping levels, GaP saturates at current densities ranging from about 1 to 10 amperes/cm^2. GaAsP is not restricted by a saturation effect and can thus be pulsed with very high current levels for brief but intense flashes of light.

Current pulsing an LED is ordinarily not practical when the device is intended for a simple indicator role. For alphanumeric readout displays, however, current pulsing is frequently a necessity. This is because drive circuitry can be significantly simplified if the individual readouts in a display are sequentially strobed. So that the display will have a satisfactory *average* brightness, the strobing pulses are at a much higher current level than that permissible for steady-state operation. This multiplexing technique works fine with GaAsP arrays, but the saturation problem makes it unsuitable for GaP arrays. More on arrays and their driving circuits will be found in Chapter 4.

From a manufacturing standpoint, the most important disadvantage of GaP is the difficulty of growing uniform, dislocation-free crystal. At the present time, commercial GaP material is consistently of poorer quality than GaAsP, and material suitable for LEDs is therefore more costly. Future prospects for GaP are good, however, and recent reports by some of the laboratories working with the material indicate better quality crystal will ultimately become available as growing techniques are improved.

While GaAs and GaP are by far the leaders in visible LED production, they receive limited competition from aluminum gallium arsenide (AlGaAs). This semiconductor is characterized by good light output at 675 nm and the ease with which very good quality junctions can be formed. The latter advantage is of great importance in a field where numerous materials with excellent electroluminescent properties are unsuited for practical light emitters due to materials problems. The excellent crystal lattice match between GaAs and AlGaAs made possible the fabrication of the first diode lasers capable of operating continuously at room temperature.

At the present time AlGaAs LEDs are not commercially available. However, the ease with which they can be fabricated may change this situation. The devices have a higher wavelength than GaAsP units, with a resultant lower luminosity efficiency. But their brightness as a function of input current density is virtually identical to that of GaAsP, because of their higher quantum efficiency.

ORANGE AND AMBER LEDs

The easiest way to obtain an orange or amber LED is to increase the GaP content in GaAsP at the expense of efficiency and brightness. While the lower wavelength (610 nm) increases the luminous efficiency of the LED from about 6 percent in the red to 50 percent, the external quantum efficiency drops from 0.5 percent to about 0.013 percent. The result is that an orange LED made of GaAsP has only about one fourth the overall visual output of an equivalent red GaAsP device. Worse, "orange" emitters sometimes appear more red than orange to many observers.

Though orange and amber LEDs suffer from low conversion efficiency, at least one company offers a commercial device. The Monsanto MV1 has a typical brightness of 500 fL at 50 mA and a forward voltage of 2.0 volts.

Laboratory developments have resulted in newer, more efficient orange LEDs that are not yet commercially available. Evidently the most practical is GaAsP that has been doped with nitrogen on the n portion of the junction. The external quantum efficiency and photometric brightness of this material is three times that of ordinary GaAsP without nitrogen doping.

An orange electroluminescent source with even more efficiency is indium gallium phosphide (InGaP). This material is presently restricted by problems in obtaining good quality crystal, but it is expected that this difficulty will be overcome as research continues. The result will be LEDs with a brightness more than three times that of conventional orange LEDs made of GaAsP. Since pn junctions can be formed in InGaP, using techniques virtually identical to those employed for commercial GaAsP emitters, the prospects of commercial InGaP LEDs are good if problems in obtaining the necessary materials can be overcome.

Thus far our discussion has been limited to LEDs with a relatively monochromatic light output. However, it is possible to obtain orange emission from Gap LEDs that emit at *two* wavelengths. These bicolor LEDs, which will be described more fully later in this chapter, employ doping levels that cause green light to be emitted from one side of the junction and red light from the other. The eye combines the green and red to get orange.

At the present time bicolor LEDs are just being offered commercially. But for a time in 1966 and 1967 Texas Instruments, Inc., offered a GaP orange emitter. The diode was labelled the PEX 1321, and its two emission wavelengths peaked at 560 nm (green) and 700 nm (red). The device had a total radiant power output of about 1.5 microwatts.

YELLOW LEDs

An interesting variety of semiconductors are available that will emit yellow light when properly stimulated. In fact, the light emission observed in silicon carbide (SiC) by Henry J. Round back in 1907 was yellow (see Fig. 1-1).

The first commercially available yellow emitters were made of SiC as well. The General Electric Company has devoted considerable effort to producing commercially feasible SiC emitters, and until recently the firm offered three devices, the SSL-1, SSL-6, and SSL-11. The SSL-1 is shown in Fig. 3-7 and the SSL-6, a glass enclosed unit, is shown in Fig. 3-8. The diodes emitted light peaking at 590 nm in the yellow and extending from the blue to the red portions of the spectrum. The spectral width of these LEDs was broader than that of GaAsP and GaP devices, and their efficiency was very low. Brightness for the SSL-1 and SSL-6 units ranged from about 25 to 65 fL at 50 mA. Also,

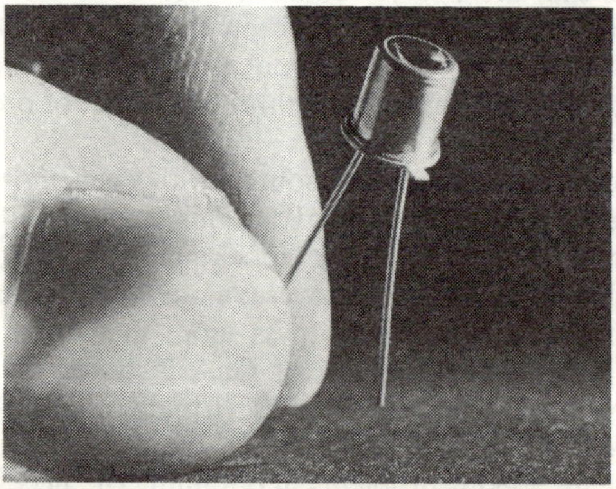

Courtesy General Electric Co.

Fig. 3-7. GE SSL-1 silicon carbide yellow emitter.

while GaAsP LEDs may have a turn-on time of only a nanosecond or so, SiC units may require 100 microseconds to reach full brightness.

For a time the GE SiC solid-state lamps offered the highest brightness of any commercial LEDs. But other semiconductors were rapidly improved and SiC diodes soon lost that distinction. Though SiC diodes show no noticeable degradation when operated under rated conditions and can withstand high operating temperatures, their relatively high price resulted in little commercial success. The reason for pricing levels several times that of other LEDs was the great difficulty encountered in growing appropriate SiC crystal. The process requires very high temperatures of up to 2500°C and results are not always repeatable.

Though GE does not offer SiC LEDs at present, new research into the original light emitting semiconductor may result in a second generation of carborundum emitters. The research will likely be slow, since considerable prior research by GE and others has not even identified all the mechanisms responsible for light emission in SiC.

Much more efficient yellow emission can be obtained from GaAsP diodes doped with nitrogen. Monsanto offers several yellow indicator LEDs as well as numeric displays made with this material. The yellow peaks at about 589 nm and the indicator LEDs have a brightness of 300 fL at 50 milliamperes. Since the brightness efficiency of yellow GaAsP emitters is higher than that of amber diodes using the same material, commercial prospects are bright.

Yellow-green emission at 570 nm can be obtained from indium gallium phosphide (InGaP). As with orange emission from this material, the efficiency is relatively high. In fact, the brightness of laboratory InGaP yellow-green LEDs is twice that of nitrogen-doped GaAsP diodes for the same current density. Unfortunately, as was noted in the discussion of orange and amber emitters, it is difficult to obtain good quality InGaP material.

Another yellow-emitting diode is the GaP unit that has been doped to simultaneously emit green and red. As with the orange GaP unit,

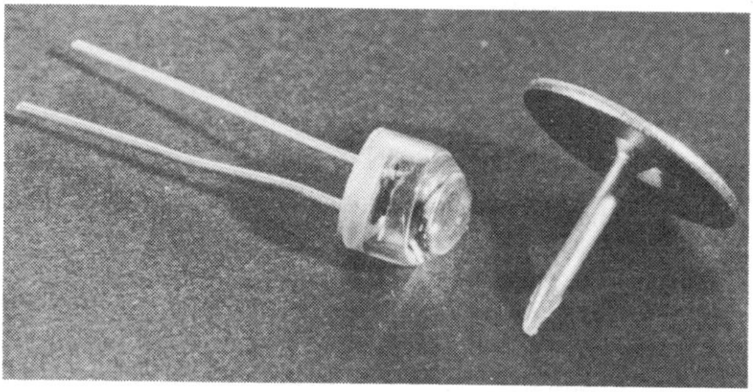

Courtesy General Electric Co.

Fig. 3-8. Glass-encased SSL-6 silicon carbide LED.

appropriate doping levels can shift the relative intensity of the green and red to give varying hues of orange and yellow. A more thorough discussion of GaP emitters of this type will be presented later.

One final kind of yellow emitter should be mentioned: the Fairchild series of avalanche-luminescent silicon planar diodes. Available from Fairchild as arrays of up to 576 discrete emitters, the devices are designed specifically for marking photographic film with symbols or data. The emission from silicon avalanche diodes does not peak at any one spot in the visible spectrum, as can be clearly seen in Fig. 3-9. The emission makes a broad swath across the entire spectrum from blue to infrared, with a peak in the near infrared.

Silicon avalanche-luminescent diodes are not practical for discrete indicators, due to their very low efficiency. The very high cost of arrays (several thousand dollars) limits their use to specialized applications at the present time.

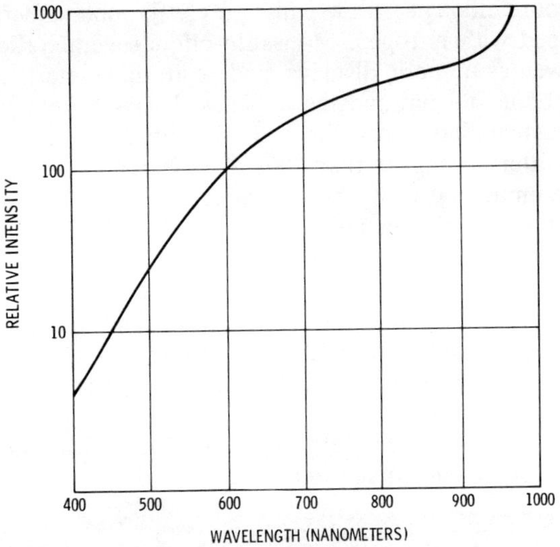

Fig. 3-9. Silicon avalanche light emitter.

GREEN LEDs

Because the sensitivity of the eye peaks in the green portion of the spectrum, there has been a concentrated effort to develop efficient and inexpensive green LEDs. This goal has finally been achieved by means of the versatile gallium phosphide (GaP) LED. Though the quality of GaP crystal still is below that of GaAsP, several companies now offer green GaP emitters. These diodes offer a brightness level higher than that of any other commercial LED for the same current density value. For example, a GaP diode identical in surface area to a GaAsP unit and operated at the same current density can be more than three times brighter than the red diode. The luminous efficiency of these diodes is the highest of any available, at 677 out of a possible 680 lumens per watt. This corresponds to a luminous efficiency of 99½ percent.

GaP green emitters do not suffer from the saturation effects that trouble their red counterparts. This is because the green emission originates from the n side of the junction, while red light comes from the p side. For this reason the prospects for green numeric displays capable of being multiplexed are excellent and already several companies offer commercial units. As with yellow numeric displays, Monsanto was the first company to market a green LED array.

At the present time GaP retains an undisputed position as the most efficient and promising green LED available. Gallium nitride devices have yielded green emission at 520 nm, but these units are experimental and restricted to laboratory use. Very low efficiency green light has been obtained from aluminum gallium phosphide (AlGaP) diodes, but these devices are also restricted to purely exploratory development.

There is an important class of green LEDs that employs an indirect technique to obtain green light. These diodes employ a GaAs:Si LED chip coated with an infrared-sensitive phosphor. When the diode is energized, the phosphor absorbs some of the infrared and converts it to a very pleasing green. The light peaks at a wavelength of about 539 nm and is therefore greener to the eye than the light from GaP

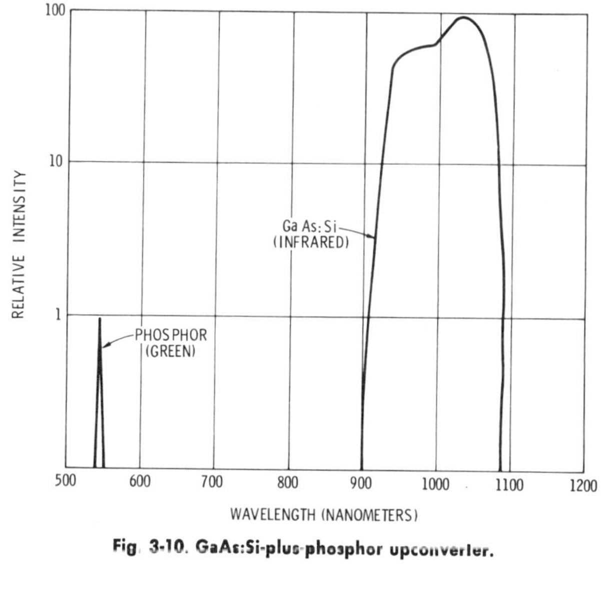

Fig. 3-10. GaAs:Si-plus-phosphor upconverter.

diodes. Since the green radiation has a higher energy than the infrared, phosphor coated units are sometimes called *upconverters*. Fig. 3-10 shows the two wavelength peaks of a typical upconverter diode.

Several companies offer phosphor upconverter LEDs, but their low efficiency and relatively high cost have hindered their commercial success.

Though they are inefficient, green upconverter diodes are desirable for some applications because of their very pleasant hue. Since the diodes emit considerable infrared as well, there are a number of control applications that can be envisioned. For example, the green radiation permits the superimposed infrared to be directed along a desired path without the need of expensive infrared viewing devices. Upconverter diodes might also prove practical for focusing lenses designed to collimate infrared LED radiation. Since the green light has a shorter wavelength, the focus would not be perfect, but the technique would permit good lens orientation.

BLUE LEDs

The availability of economical and efficient blue LEDs would nicely round out the color range of visible LEDs, but at present blue emission has only been observed under laboratory conditions in a very few semiconductors. The most common material now being researched is gallium nitride (GaN). The material has a high resistance, so electrical efficiency is low. And since the blue emission at 440 nm is well down on the photopic luminosity curve with a luminous efficiency of only about 3 percent, the GaN devices suffer from very poor visual efficiency.

The major difficulty with GaN is the problem in obtaining low resistance p-type material. Because of the high internal resistance of laboratory blue emitters a relatively high operating voltage is required. The availability of low-resistivity p-type material could likely make GaN one of the most efficient sources of visible light in the blue, green, and yellow, so research into the material is continuing.

It should be noted that SiC is also capable of emitting blue light. In fact, the emission from commercial SiC LEDs includes some blue. But as was discussed under yellow emitters, there are many drawbacks to SiC as a practical light source.

MULTIPLE-COLOR LEDs

For some time it has been known that controlled doping can cause GaP diodes to emit green and red light simultaneously. Since red emission from GaP saturates at relatively low-current densities, the green emission can be made to predominate by operating the diode at higher

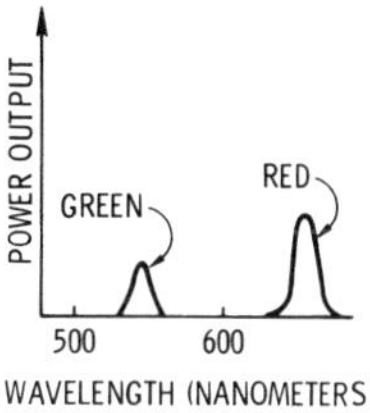

(A) Red (low current).

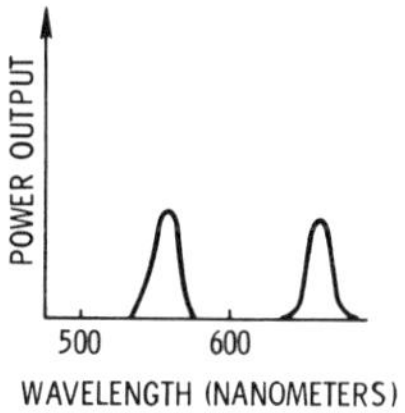

(B) Yellow (medium current).

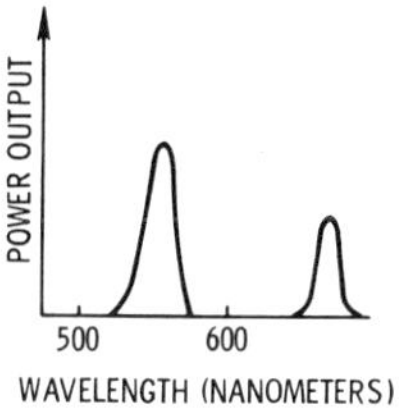

(C) Green (high current).

Fig. 3-11. Spectral output of multiple-color LED.

currents. The result is a diode that appears red at low current and green at high current. In between, the diode will appear as varying hues of orange and yellow. The effect is summarized in Fig. 3-11.

Several companies are now researching these multiple-color emitters, and it is expected that units will ultimately become commercially available. Potential applications for the devices as circuit-status indicators and even as analog-voltage indicators should be interesting possibilities for circuit designers and experimenters. It is conceivable that multiple-color LEDs could even be used to send multiplexed signals through a single optical-fiber communications link.

Another technique to achieve multiple-color emission is to enclose two or more LED chips in a single package. If one is GaAsP (red) and the other is GaP (green), for example, various shades of red, orange, yellow, and green can be made to appear by merely adjusting the current to each diode. Several companies market multiple-color LEDs made in this fashion.

SUMMARIZING LED COLORS

Much of the information contained in this chapter is summarized in Fig. 3-12, a diagram showing the wavelength spread of several common LED materials superimposed on the photopic luminosity curve. The figure illustrates the relative spectral sensitivity of the human eye to typical LEDs. The difference between green and red emitters is quite striking.

Table 3-1. Characteristics of Common Visible LEDs

Semiconductor	Peak Wavelength (nm)	Color	Ext Quantum Efficiency (%)	B/J (see text)
GaP, N	550	Green	0.1	470
SiC	590	Yellow	0.003	10
GaAsP	610	Amber	0.013	35
GaAsP	660	Red	0.5	145
GaP:Zn,O	690	Red	3	350

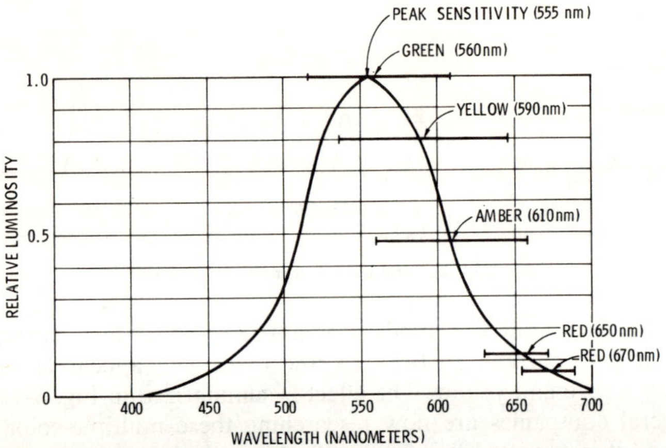

Fig. 3-12. Wavelength spread of common visible LEDs.

Table 3-1 summarizes the characteristics of several LED materials that emit visible light. The most important specification in the table is B/J, brightness in footlamberts divided by the current in amperes per cm^2. This specification provides an excellent comparison between diodes that emit at different wavelengths.

COMMERCIAL VISIBLE LEDs

The success of visible-light-emitting diodes has prompted the entry of dozens of semiconductor and incandescent pilot lamp manufacturers into what has become a highly competitive field. Since solid-state lamps are more rugged, easier to make, and have much longer lifetimes than their incandescent counterparts, design engineers are employing them in many circuit applications. LEDs are frequently used for status indicators and numeric readouts on modern test equipment.

The biggest market yet for visible LEDs is the booming miniature electronic calculator market. The demand for calculator displays is such that some manufacturers have experienced delays in obtaining the GaAsP substrates required for making the displays. As the supply catches up with the demand, lower costs will result for both discrete indicator LEDs and displays.

Fig. 3-13 shows an assortment of common visible LEDs. While LEDs are available in dozens of configurations, the most popular for low-cost indicator applications are those completely encapsulated in plastic. The units are very economical and relatively rugged, and the plastic significantly enhances optical output by providing a better index-of-refraction match to the light-emitting crystal than air. An important advantage of plastic LEDs is that they can be fabricated automatically by means of equipment developed for automated production

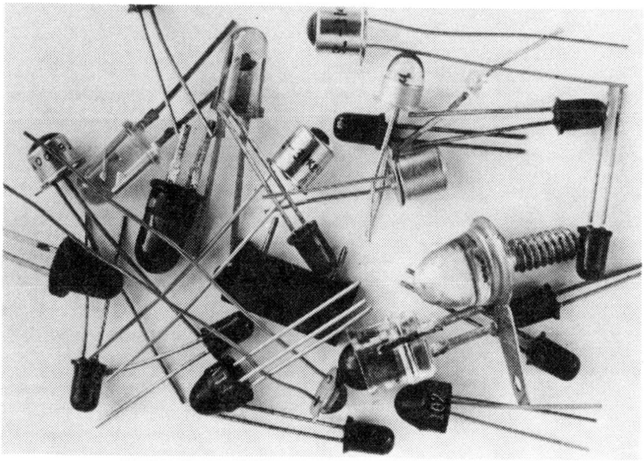

Fig. 3-13. Assortment of commercial visible LEDs.

of plastic-encased transistors. Fig. 3-14 is an enlargement of one of the most common plastic emitters, the Litronix Red-Lit 50. A second-source version of the Monsanto MV50, this emitter is one of the least expensive available. Note the internal construction of the diode (active surface, metal contact region, wire electrode, etc.) in the photograph.

Another popular configuration for all-plastic LEDs is shown in Fig. 3-15. These General Electric diodes are sturdier and easier to handle than the MV50 style units and are therefore preferred for use as panel lights. A similar type of plastic LED is the Hewlett-Packard MINI-LED shown in Fig. 3-16.

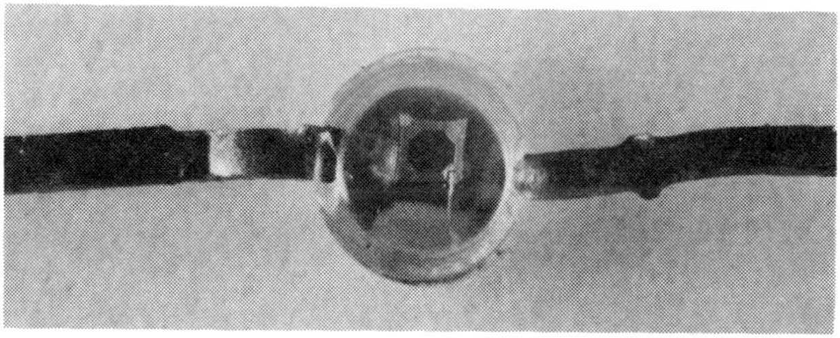

Fig. 3-14. Litronix Red-Lit 50 miniature GaAsP LED.

The next most popular visible LED package consists of a metal base topped by a plastic blob that serves as both a lens and a protection medium for the chip and contact leads. These diodes are usually somewhat more costly than their all-plastic counterparts, but they offer advantages of mechanical rigidity and heat sinking. Because of the increased metal over that of external contact leads alone, the metal-based diodes can operate at much higher current levels.

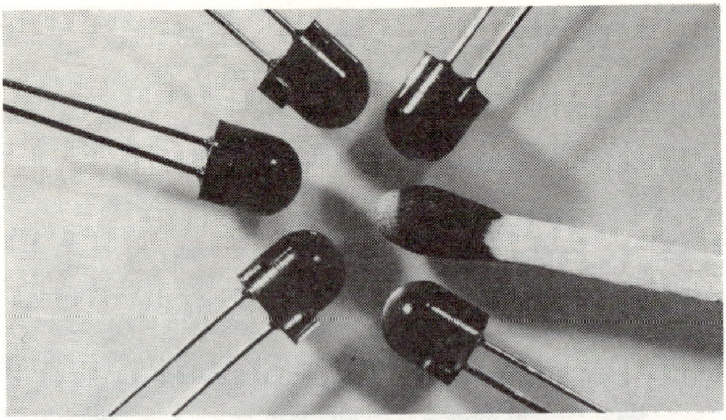

Courtesy General Electric Co.

Fig. 3-15. SSL-212 miniature red-emitting LEDs (GaP).

An extreme version of a plastic-lens, metal-base diode is Monsanto's MV4H, the "Big Red." This high-power red emitter is made with GaAsP and emits a full milliwatt at 500-mA forward current. The photometric brightness at this power output is an impressive 2500 fL. With a proper heat sink the device will emit 5000 fL at 1 ampere. In Fig. 3-17 a girl is shown admiring the appearance of this big diode.

For applications where a hermetic seal is essential, LEDs are available in conventional transistor cans with a glass lens. These diodes can operate at high current levels because of increased heat sinking, but their viewing angle is restricted. Also, the efficiency with which light generated within the LED chip is removed to the outside is decreased by the opaque metal wall of the can.

Arrays of visible LEDs designed for alphanumeric arrays are available in a variety of plastic and glass packages. Commercial devices will be described in detail in Chapter 4.

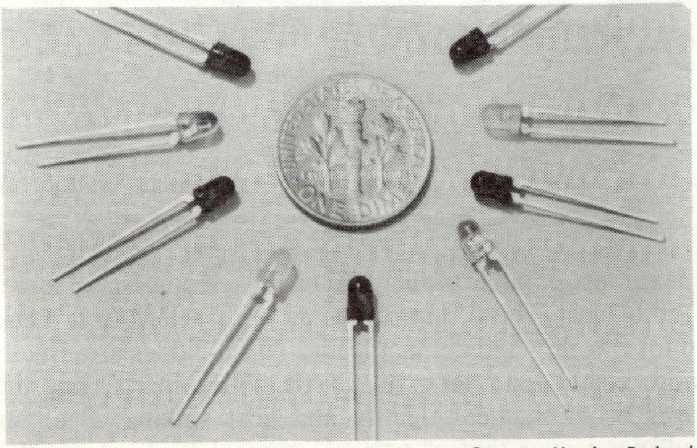

Courtesy Hewlett-Packard

Fig. 3-16. Red, clear, and diffused-lens versions of HP Model 5082-4480 Mini-LEDs.

Fig. 3-17. Monsanto MV-4H high-power red LED.

EVALUATING VISIBLE LEDs

The confusing aspects of photometry have resulted in what is no doubt the most controversial area of specifying LED output. Perhaps half a dozen different photometric units are employed in rating the visible output of relatively similar devices. In addition to rating output in terms of lumens, manufacturers also employ units of luminance or brightness (footlambert) and luminous intensity (candela).

Ideally, each manufacturer would rate its LEDs in terms of optical flux, brightness, and intensity. Optical output should be given as a function of *current density* as well as a fixed forward current. Current density is important because it provides a means to compare different diodes. It is hard to tell much about two diodes that each produce 700 fL at 10-mA forward current, but when we are told that one diode has a smaller current density—hence smaller chip size—we automatically know that it is more efficient. And since it uses a smaller chip, it is potentially less expensive.

One reason that manufacturers do not use consistent photometric ratings is the dramatic difference which results for point- and area-source diodes. In the former case the diode is encapsulated in a clear epoxy or provided with a transparent glass lens. Area-source diodes, however, employ a diffusing lens, usually the color of the emitted light or diffuse white, to spread the flux over a wider viewing area. Consequently, area sources do not have the point intensity of the devices without diffusing or spreading optics.

The different rating techniques can be nicely compared for an identical diode available with or without a diffusing lens. Consider the Fairchild FLV100 and FLV101 gallium arsenide phosphide, red-emitting LEDs, for example. Fairchild has been generous enough to provide a data sheet that rates these visible LEDs with three photometric units and gives a conversion table that can be used for extrapolating a few others.

With the exception of a red plastic diffusing lens on the FLV101, both LEDs are identical. Yet the luminance or brightness of the FLV-100 (1500 fL) is substantially greater than that of the FLV101 (18 fL). The reason for the dramatic difference is that the apparent optical source of the diffused lens diode is much larger than that of the clear-lens, point-source diode. While both LEDs emit approximately the same optical flux (the diffuse red lens absorbs about 7 percent of the light output of the FLV101), the uninitiated customer relying on data sheets alone would probably choose the FLV100 over the FLV101, based on its superior luminance rating.

Actually, the larger source size of the FLV101 makes it a more practical LED for many indicator applications. The light from a point-source LED is so intense and originates from such a small spot that viewing can be uncomfortable in some situations. The addition of a diffuse lens spreads the light over a wider area, increases the viewing angle, and provides a more comfortable source of light.

There is another important benefit to a diffusing lens the same color as the LED output wavelength. In a report on one of the first commercial red LEDs, it was noted that the Monsanto MV10B3 could be seen for 500 feet at night while drawing only about 10 milliamperes from a 1.5-volt flashlight cell. In sunlight, however, even at higher current levels and while viewed from only a few inches, "the diode washes out and all you'll see if you squint right at the die is a dull red chip."

The problem here is contrast. The MV10B3, with its clear epoxy lens, permits ambient light to be reflected from the shiny internal parts of the diode such as the metal header, the connection lead, the evaporated metal contact regions on the chip, and the chip itself. Since the power density of the ambient light may easily equal or exceed the emission from the LED, the diode output is lost to the bright reflections of ambient light.

When the chip is immersed in a plastic containing a dye transparent to the LEDs wavelength, only ambient light with the same color as the LED output can enter the epoxy. This is a much narrower spectrum than with clear epoxy, and the LED output generally appears much brighter. The result is that the dyed epoxy acts as a contrast filter and the LED can be easily viewed in bright ambient light.

Another technique of enhancing contrast is the use of a polarizing filter. Since the ambient light reflected from the internal structures of the diode becomes polarized, the filter prevents much of it from exiting while presenting little resistance to its initial entry. The polarizing only blocks a small part of the LED light.

Frequently LED makers provide appropriately colored epoxy lenses on their diodes (such as the FLV101). But clear-lensed diodes can be provided with contrast enhancement by means of an external red or circular polarizing filter. Several manufacturers offer the filters as optional items.

This discussion on evaluating visible LEDs can be best summarized with a simple but practical rule: *look before buying.* The photometric terms and data provided by the manufacturer on data and specification sheets can provide some indication of relative brightness but only when comparing similarly fabricated LEDs. The commonly used brightness term, the footlambert, is of little or no value when comparing two diodes if one has a diffusing lens.

Manufacturers frequently supply electronic parts distributors and sales representatives with elaborate display boards containing a dozen or more brightly lit LEDs. By examining such a display board, even the experimenter contemplating purchase of one or two LEDs can make a better choice if the devices are to be used as visual indicators. Manufacturers of equipment should never buy in large quantity without obtaining some representative samples for in-circuit evaluation. Most manufacturers are willing to supply small quantities free to potential large-volume customers. Since many manufacturers offer LEDs identical in light output and mounting configuration to those made by competitors, the large-volume customer would be prudent to consider their purchase. In this manner the purchaser judiciously protects himself in the event the original supplier cannot keep up with production or if any of a variety of manufacturing hazards eliminate production completely. This technique is commonly used in the purchase of large quantities of components and is called *second sourcing.*

Another important consideration in specifying visible LEDs is optical overkill. A high degree of visibility is usually required for panel light applications, but a relatively low output can usually be tolerated in, for example, circuit-board troubleshooting and status indicators. Most visible-LED makers offer miniature plastic encapsulated units selling for less than 50¢ in volume quantities. Requiring only a milliampere or two for a luminance of up to 50 or more footlamberts, these tiny, plastic encapsulated LEDs put out a great deal of light for use as circuit-board status indicators.

Finally, remember the point-source visible LEDs for their particular advantages. While the combination of diffusing lens and red filter enhances visibility, other applications may require the high intensity of a point source. For example, with external optics it is relatively simple to collimate the light from a point-source LED into a narrow beam. And in many applications where the LED is to be modulated with a signal or used in a control circuit, the point-source diode is usually the best choice.

CHAPTER 4

LED Displays and Arrays

Light emitting diodes are almost ideally suited for fabrication into arrays. In the case of visible LEDs, arrays can be built that display the digits 0 to 9, letters of the alphabet, and practically any conceivable symbol. There are even arrays that display Japanese characters. Displays of this type are finding wide application in miniature electronic calculators, digital clocks and watches, and, in the near future, automobiles.

By far the biggest application for visible LEDs is alphanumeric displays. A typical display requires a minimum of 8 individual LEDs, and many use from 14 to 36 or more.

The important advantages of LED visible arrays are their compact size, low operating voltage, and mechanical rigidity. Formerly, gas discharge tubes (such as the Nixie) were the only display devices available, but their high voltage requirement makes them generally unsuitable for many transistor circuit applications. LED displays do suffer from a disadvantage of relatively high current consumption. The current consumed by a single diode is low, perhaps a few milliamps or less. But when a substantial number of diodes in a typical display are illuminated, the current consumption can easily rise to a hundred or more milliamperes.

Fortunately, not all diodes in a display are on at any one time, so average current consumption is not always high. And some displays incorporate blanking circuits to turn the LEDs off at intervals in order to save power.

Arrays of infrared LEDs can also be fabricated (see Fig. 2-14). An important use for such devices is in high-power covert infrared illumination systems for observing scenes in total darkness. In this application an infrared image converter supplies the electronic "eye" necessary to view the scene. Infrared diode arrays also find use in rocket-tracking beacons, laser pump sources, and other applications.

This chapter will be concerned with visible arrays. Besides covering the various kinds of arrays, electronic techniques for operating them will be described in detail.

DIGITAL DISPLAYS

Several interesting shortcuts have been devised to permit the digits 0 to 9 to be displayed with only a limited number of optical elements. All depend on dividing the various digits into segments or dots that can be illuminated in various combinations to give the numbers desired. The most common digital display is the seven-segment unit.

Fig. 4-1 shows the layout for a typical seven-segment display and how the segments are illuminated to give the digits 0 to 9. Most seven-segment displays incorporate a single LED as a decimal point, and displays are available with either a left or right decimal point. Special-purpose displays may incorporate other additional LEDs. For example, a display intended for use in a digital clock may incorporate a colon to denote the spacing between hours, minutes, and seconds.

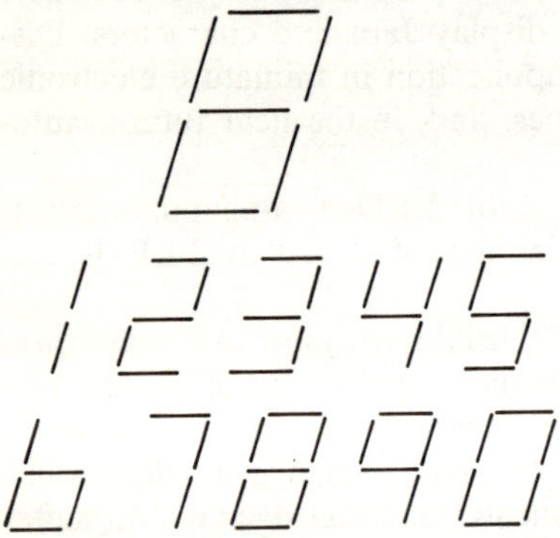

Fig. 4-1. Seven-segment numeric display.

Seven-segment displays are available in a wide range of sizes, configurations, and prices. Fig. 4-2 shows a representative seven-segment display made by General Electric. Some displays incorporate integrated circuitry designed to permit simplified addressing of the display and other special-purpose functions, and some of these will be described later in this chapter. At present, one of the most popular displays is the Monsanto MAN-1. This unit has a 0.27-inch character height and can be easily read from a distance of several yards, so long as the unit is not illuminated by bright ambient light. The brightness of each segment is about 350 footlamberts at 20 mA. With all segments illuminated, the display consumes 480 mW.

A Litronix equivalent to the MAN-1 display is shown in Fig. 4-3. The display is made from a lead frame assembly that has been cast within a plastic material. The active elements of the display consist of seven pairs of bar-shaped GaAsP LEDs and a single square LED that serves as a decimal indicator. Examination of an individual LED bar within the display through a 50-power microscope will reveal a metallized grid pattern that serves as the electrical contact point to the n region of the diode. A tiny wire electrode is connected from this metallized layer to the p region of the adjacent diode so that each pair of LEDs is in series. Connections to each pair of LEDs and the decimal indicator are made via eleven external pins.

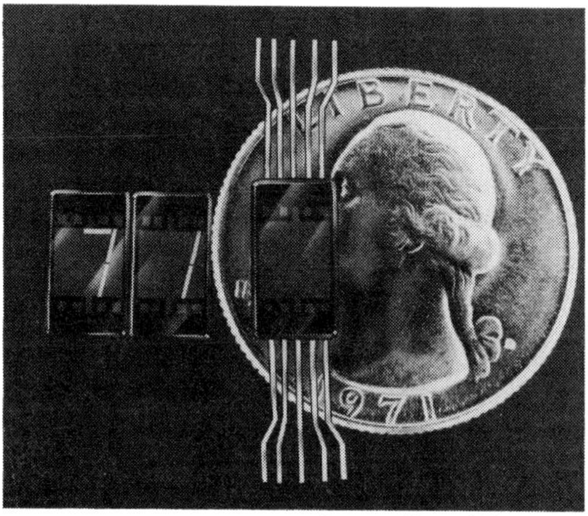

Courtesy General Electric Co.

Fig. 4-2. SSL-140 and SSL-190 monolithic seven-segment displays.

The MAN-1 display can be easily inserted into a matching socket or soldered directly into a circuit board. Some other displays are much smaller and are available as modules of three or more individual readout units. The small displays are of two general types: hybrid and monolithic. Hybrid units employ separate LED bars or circles. Monolithic units consist of a GaAsP substrate upon which a metallized seven-segment pattern has been affixed by photolithographic techniques. While the substrate is one large chip of semiconductor, only

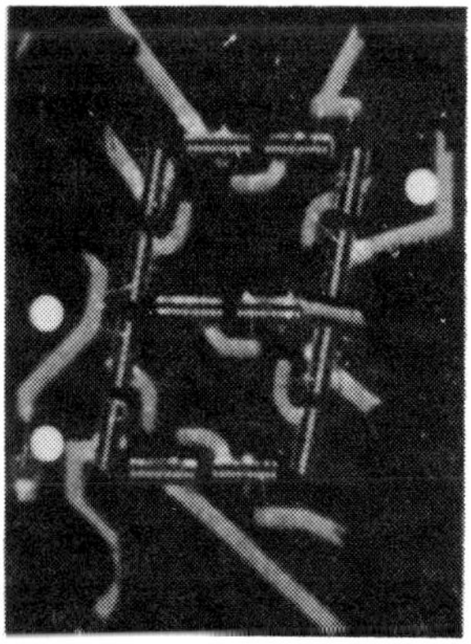

Fig. 4-3. Litronix Data-Lit 10 LED display.

areas of the crystal in the vicinity of an electrode are illuminated when current is applied. Fig. 4-4 shows three Monsanto monolithic displays mounted to a common base.

Hybrid displays require far less semiconductor material than the monolithic units, but they are more tedious to assemble. Proper fabrication requires that each of the seven pairs of diodes be correctly aligned and then soldered in position. Monolithic displays are much easier to assemble, since the seven segment pattern can be formed during a single processing step, using conventional photolithographic techniques. But monolithic displays waste GaAsP since most of the substrate is not optically active. For this reason the higher assembly cost of hybrid displays is balanced by the higher materials cost of monolithic units. Fig. 4-5 compares the construction of a portion of a hybrid and monolithic display.

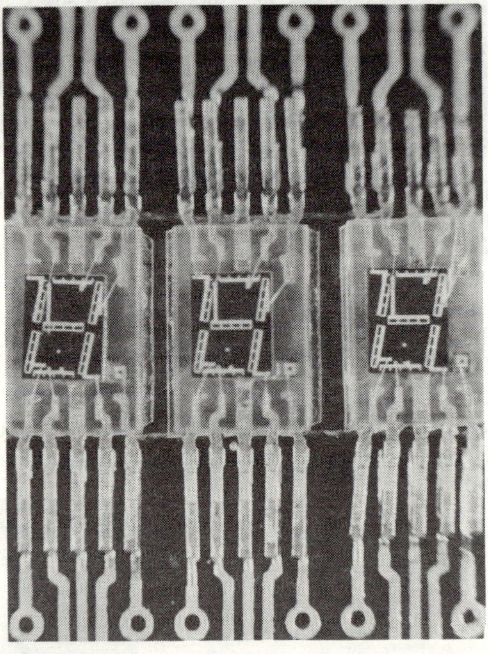

Fig. 4-4. Monsanto MPC2 monolithic display array.

Monolithic displays are very small, with typical character heights ranging from 0.1 to 0.115 inch, or less than half that of the MAN-1. Brightness can be as high as 400 fL at 10 mA with the devices being visible at 1 mA. Monolithic displays offer advantages of compact spacing, and when material costs are lowered, their prices should fall significantly.

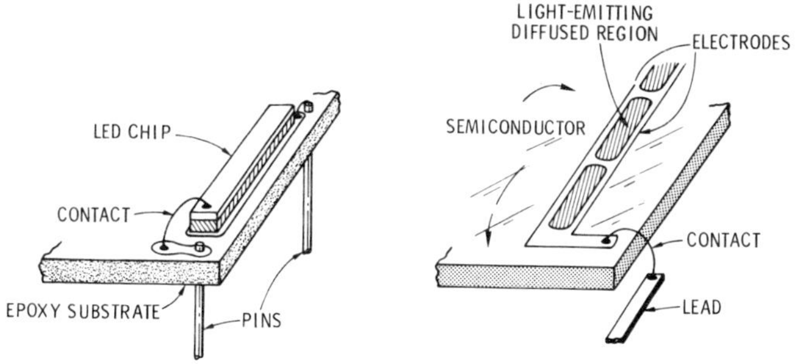

(A) Hybrid. (B) Monolithic.

Fig. 4-5. Hybrid and monolithic display construction.

OPERATING DIGITAL DISPLAYS

To generate a digit on a seven-segment display, appropriate electronic circuitry is required. The most elementary addressing circuit would use a series of mechanical switches as shown in Fig. 4-6. To form the digit 5, for example, switches 1,2,4,5, and 6 are depressed. The resistors serve to limit current to the individual segments in the display.

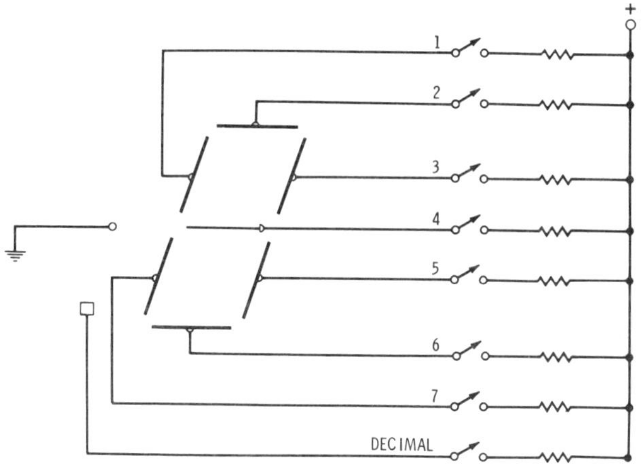

Fig. 4-6. Switching network for seven-segment display.

In most digital electronic circuits the output is in the form of binary-coded decimal (BCD), and it is convenient to employ an electronic decoding network to change the BCD into a seven-segment digital number. Integrated circuits are available for performing this function and they are referred to as BCD to seven-segment decoder/drivers.

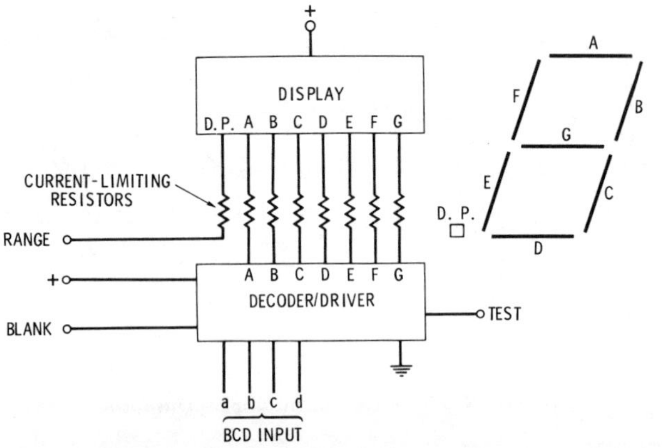

Fig. 4-7. Basic BCD seven-segment block diagram.

As will be noted later in this chapter, some numeric displays incorporate integral BCD decoding logic.

A block diagram for a typical BCD decoder/driver is shown in Fig. 4-7. In operation, BCD data is fed into the decoder input. The IC converts the BCD into a seven-segment format and the appropriate segments of the display are turned on. Since the individual segments in a display usually require no more than 10 mA for proper operation, it is possible to fabricate the driving transistors directly onto the decoder chip. Besides the decoder/driver and display, the only additional components required are eight current-limiting resistors for the LEDs.

Note that the decimal indicator is not connected to the decoder/driver. Instead it serves as a range indicator and is driven separately. Note also that the decoder/driver incorporates a lamp-test connection and provisions for blanking the display. The lamp test connection

Table 4-1. Truth Table for Seven-Segment Display

BCD input				Output to segments							Display
d	c	b	a	A	B	C	D	E	F	G	
0	0	0	0	0	0	0	0	0	0	1	0
0	0	0	1	1	0	0	1	1	1	1	1
0	0	1	0	0	0	1	0	0	1	0	2
0	0	1	1	0	0	0	0	1	1	0	3
0	1	0	0	1	0	0	1	1	0	0	4
0	1	0	1	0	1	0	0	1	0	0	5
0	1	1	0	1	1	0	0	0	0	0	6
0	1	1	1	0	0	0	1	1	1	1	7
1	0	0	0	0	0	0	0	0	0	0	8
1	0	0	1	0	0	0	1	1	0	0	9

enables all the display segments to be activated with a single voltage connection to permit all the display to be quickly checked. A nonoperating display segment can change a seven to a one or an eight to a zero, and cause serious readout errors.

Table 4-1 shows the various BCD input conditions required to generate each of the digits 0 to 9. *Truth tables* such as this are frequently employed in describing the operation of digital logic circuits.

MULTIPLEXING DIGITAL DISPLAYS

It is possible to arrange a series of displays and decoder/drivers so that multidigit numbers can be displayed. When the display goes beyond several digits, however, the cost of individual decoder/drivers can become higher than the components required for a novel form of display operation called *multiplexing* or *strobing*. This technique makes use of but one decoder/driver, which is sequentially time shared with each of the display units, one at a time. All of the respective segments of the display devices are wired in parallel so only a connection to the anode (or cathode) of each position is required to activate a single digit in the display. A clock circuit shifts the character-enable voltage from one digit to the next, while a single BCD seven-segment decoder provides the appropriate segment connections.

If a display is multiplexed without increasing the current to each LED segment, it will appear very dim and may not be possible to read. This situation is easily remedied by increasing the strobing current over that allowed for dc operation. If the average multiplexed current equals the dc current, the display will appear equally bright for both conditions. As when operating discrete LEDs in a pulse mode, the maximum allowable duty cycle must not be exceeded or the diodes in a display will be degraded and possibly destroyed. A typical display can be operated at a peak current of 100 mA per segment with a pulse width of 50 to 200 microseconds and a duty cycle of 20 percent. By varying the pulse width, the display brightness can be varied, but precautions must be taken to avoid exceeding the allowable duty cycle.

Fig. 4-8 shows the LED display, driving transistors, and series resistors for a Sharp ELSI MINI miniature electronic calculator. The use of multiplexing makes the display assembly very compact.

The strobing technique works well with GaAsP red and yellow diodes and GaP green emitters. But because of the inherent saturation problem of GaP red LEDs, strobing is not practical with displays made with this material. Opcoa does manufacture GaP red displays with more efficiency than GaAsP units. While they cannot be strobed, there are certain applications where separate decoder/drivers are as economical as multiplexing and where the GaP units may then be preferred as superior displays.

A few comments about the subjective viewing of multiplexed displays should be made. First, due to a slight nonlinearity in the curve of forward current versus brightness for GaAsP LEDs at very low current levels, a strobed display may actually appear brighter than a dc display with identical average current. This is because the strobing technique operates the LEDs in a more efficient mode, the linear portion of the current-versus-brightness curve. The increase in apparent brightness may also be due to the fact that the human eye is capable of responding to part of a fast pulse. Since the eye is also an integration device in that it tends to average the light being observed, the brightness for a pulsed display may appear higher than for a dc unit operated at an identical average current.

Fig. 4-8. Sharp ELSI MINI miniature electronic calculator display board and circuitry.

A second point about subjective viewing of a display concerns the strobing frequency. To the eye, a light-pulse rate of about 16 repetitions per second appears as a continuous glow, though this rate may vary somewhat with different subjects and viewing conditions. A multiplexed display must energize each digit at least twenty times per second to avoid any noticeable flicker from the display units. For situations where the display may be exposed to mechanical vibrations, much higher strobing rates (e.g., 100 Hz) must be employed to avoid flicker. This is because a movement of the display will cause individual digits to be energized at various points in space, resulting in a noticeable flicker.

It is particularly important to employ a fast strobing rate in digital wrist watches, pocket calculators, and other portable equipment. Under normal conditions a 100-Hz strobe rate will appear flicker free, but even a rate of 120 Hz employed in one firm's pocket calculator flickers noticeably if the machine is rapidly moved. Of course, the device is not moved fast enough during normal operation to cause noticeable flicker.

The flicker one sees when a strobed display is rapidly moved provides a simple method of determining if the display is operated in a multiplex mode.

ALPHANUMERIC DISPLAYS

The standard seven-segment display can produce the letters, A,C,E,F,H,J,L,P, and U as well as each of the digits. For applications where more alphabet characters are desired, displays containing thirty-five or more individual LEDs in an X-Y array are employed. The most common configuration is the 5 × 7 array with decimal indicator. Fig. 4-9 is a closeup of a Texas Instruments 5 × 7 dot matrix array. A display containing 5 individual 5 × 7 dot matrices is shown in Fig. 4-10. This Hewlett-Packard display contains 175 individual LED chips.

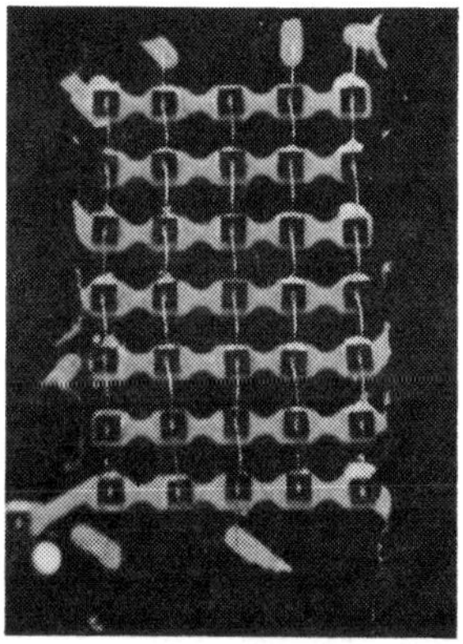

Fig. 4-9. Texas Instruments TIL305 5 X 7 LED array.

It is immediately obvious that a major drawback to 5 × 7 alphanumeric arrays is their cost. Prices are typically two or more times that of similar size, seven-segment displays. Nevertheless, some applications require the flexibility of multiple-character generation, and the added cost of the 5 × 7 array is frequently justifiable. Fig. 4-11 dramatically illustrates the flexibility of a display employing 5 × 7 alphanumeric arrays. This display on a Hewlett-Packard HP-9820 programmable electronic calculator incorporates a total of 560 individual GaAsP LEDs.

In addition to the alphabet and the digits 0 to 9, a 5 × 7 matrix display can generate more than 64 symbols and characters. There are several ways to generate characters with external logic and a read-only memory (ROM) containing the necessary coding information for generating individual symbols. The easiest way to form alphanumeric characters is by multiplexing. Appropriate external circuitry generates the necessary signals for each LED in the array while a scanning circuit connects the character generator to the array, one row or column at a time. As the row or column is connected, the appropriate LEDs are activated for a small fraction of a second, and the next row or column is connected. When the diodes are strobed at a rate faster than a hundred times per second, the slow response of the human eye will integrate the rapidly flashing LEDs to give a nice, uniform appearance. If the scanning rate is too slow the diodes will appear to flicker.

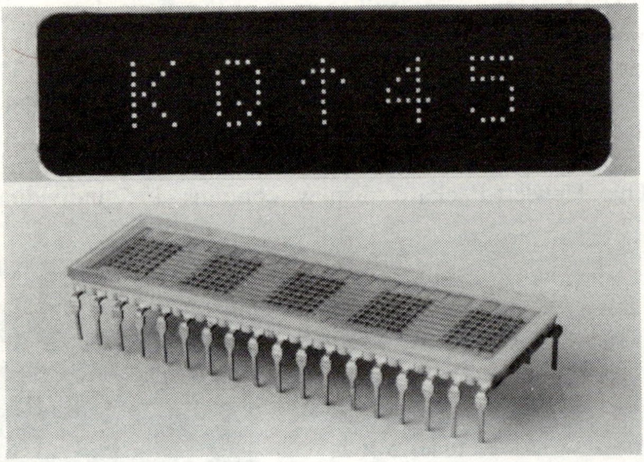

Courtesy Hewlett-Packard

Fig. 4-10. HP 5082-7102 five-character 5 × 7 LED arrays.

There are two basic strobing techniques for 5 × 7 arrays. *Vertical multiplexing* is obtained when the horizontal rows of five LEDs are sequentially scanned. *Horizontal multiplexing* is obtained when the vertical columns of seven LEDs are sequentially scanned. Vertical scanning is generally preferred when several displays will be operated with one read-only memory, since fewer external components are required.

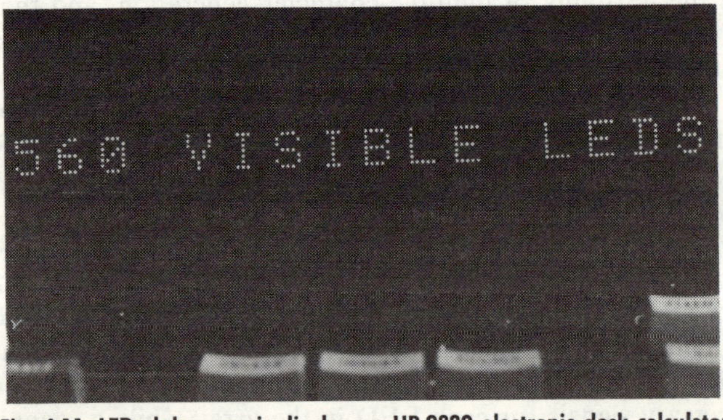

Fig. 4-11. LED alphanumeric display on HP-9820 electronic desk calculator.

POLARITY AND OVERFLOW DISPLAYS

Most LED manufacturers offer several symbol configurations designed to supplement both seven-segment digital displays and 5 × 7 dot matrix arrays. These symbol displays generally contain only a limited number of LEDs, as they are intended only to generate simple symbols such as "+", "–", and the digit "1". A digital clock might incorporate a "1" array, while a pocket calculator might use a combined "+' and "–" readout. Since these displays utilize less semiconductor material than their digital and alphanumeric counterparts, they are potentially less costly. But since a typical readout may employ as many as ten or twelve conventional displays and only one polarity and overflow display, it is not possible to get the large-quantity price break that comes with purchasing in volume. Therefore, the polarity and overflow indicator may end up costing the same as the more complicated alphanumeric display.

There are several variations of symbol displays. One is the colon, two LEDs designed to separate the hours, minutes, and seconds on a digital-clock readout. One of these special purpose displays is shown in Fig. 4-12.

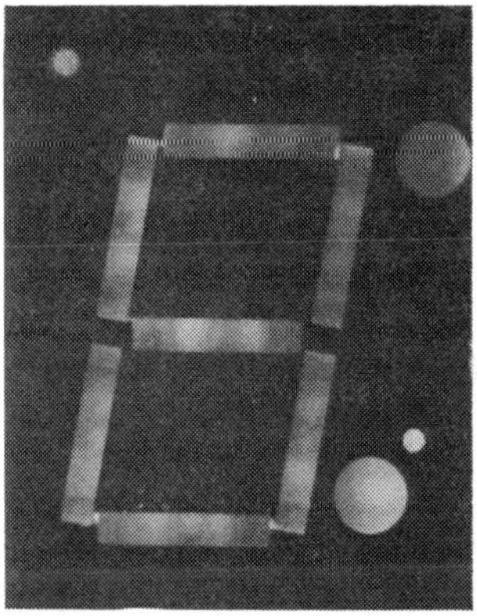

Fig. 4-12. Litronix seven-segment display with colon for digital clocks.

DISPLAYS WITH LOGIC

The most significant advance in LED readouts is the incorporation of self-contained logic circuitry. The integral logic technique used by Hewlett-Packard in the numeric display module of Fig. 4-13 employs conventional BCD logic to generate the digits 0 to 9. A separate input is provided for a decimal indicator. Since the display incorporates a 4 × 7 dot matrix, the digits are more stylized than with the more limited seven-segment approach. Fig. 4-14 shows a newer Hewlett-Packard display with integral logic, the 5082-7300, in the illuminated configuration.

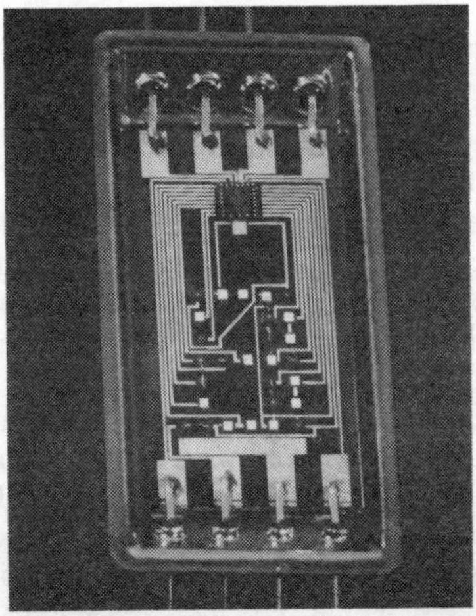

Courtesy Hewlett-Packard

Fig. 4-13. HP 5082-7000 numeric display with integral decoding and driving logic.

Texas Instruments incorporates a self-contained TTL MSI chip and counter in their TIL306 numeric display. This versatile device is virtually a complete subsystem in itself, and it is possible to employ half a dozen TIL306s in a frequency counter a small fraction of the size of its conventional counterpart. The TIL306 incorporates a seven-segment display with a character height of 0.27 inch and either a left or right decimal indicator. The integral MOS integrated circuitry contains a BCD counter, a four-bit latch, and a decoder/driver. The device has only sixteen pins.

OTHER DISPLAYS

LED displays are by no means limited to the alphanumeric configurations just described. Several years ago, for example, some engineers

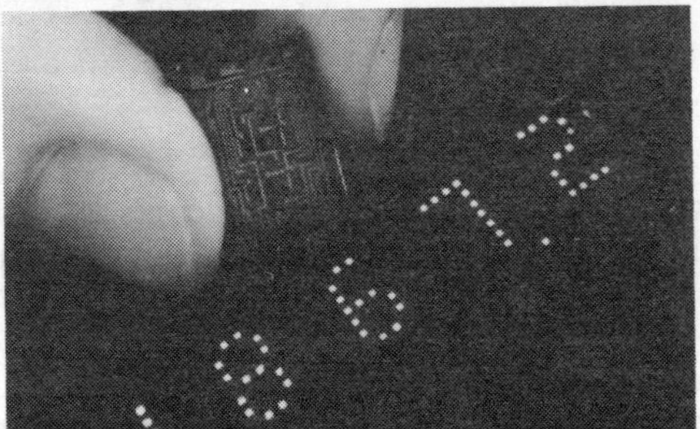

Courtesy Hewlett-Packard

Fig. 4-14. HP 5082-7300 numeric display with self-contained decoder/driver.

at Texas Instruments built a clock with a LED display consisting of a circular ring of discrete emitters around the face of the clock. Individual diodes were activated to signify the time, and the time was read from the clock as if it had hands.

There are other ways in which LEDs can be used to provide a display. One of the simplest is the use of a multiple-color LED as an analog voltage meter. General Electric has devised a 96-element analog display in a bar graph format. The display is designed for aircraft use and can be viewed in 10,000 foot candles of ambient illumination. In operation, all LEDs below the number indicated on the display are illuminated. A small numeric array above the linear bar graph provides a digital reading of the analog indication.

GE's bar graph concept could be easily modified to reduce current consumption by having only one LED illuminated at any one time. However the display might not be as readable. Such a display would be well suited for use in a chart recorder using light-sensitive paper. Unlike mechanical recorders, the LED display would not have to be moved across the paper and would therefore be much faster.

VIEWING CONSIDERATIONS

Most of the semiconductor compounds that have been used in individual visible-light emitters have been employed in visible arrays. As with discrete emitters, GaAsP is presently the most common compound. A few companies make GaP arrays; however, though they are more efficient than GaAsP, their low saturation current makes them unsuited for use in displays where each readout element is strobed sequentially.

Many people find green to be a more pleasant color than the red used in almost all LED displays, so there is a great deal of effort to increase the quality and reduce the cost of GaP green LED material. At present only a few firms offer green arrays, but recent improvements in GaP production should increase the availability of this material. One firm, Monsanto, offers a GaAsP *yellow* array.

As with discrete visible LEDs, an important consideration for displays is visibility. Brightness alone is not necessarily the only evaluating consideration. The character height and width, contrast ratio, and ambient lighting conditions dramatically affect both visibility and readability of a display. Of particular importance is display uniformity. Displays with two individual LED chips per segment are very inexpensive, but they are very difficult to read. The small point sources of light make viewing uncomfortable at best. Continuous segments which employ bar-shaped LED chips are much easier to view.

Viewing angle is another important consideration. As an example of the subjective nature of visible displays, it can be noted that many units have characters intentionally tilted about 10° from the vertical. This format enhances readability and makes a display more "pleasant."

Several ingenious techniques have been devised to enhance display readability. Opcoa uses an integral reflector to increase the light output of its GaP displays. Sony has developed a display that employs GaP segments mounted edgewise so that the viewer sees the junction directly. This improves the brightness of the display.

The Sony technique permits both red- and green-emitting GaP segments to be mounted adjacent to one another. By controlling the current through the segments, it is possible to obtain colors ranging from red to green. Conductive epoxy is used to make electrical contact to the segments as well as to hold them in place.

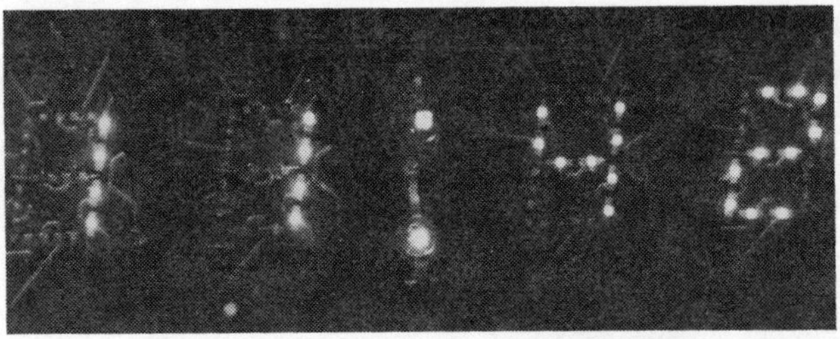

Fig. 4-15. Poor contrast LED digital clock display.

Because of the very subjective nature of visible light sources, it is frequently advantageous to employ external optical techniques to enhance the appearance of LED displays. A major consideration is contrast improvement. As with discrete indicator LEDs, ambient light reflecting from shiny materials within a display package can all but wash out the display itself. The result, as shown in Fig. 4-15, is poor contrast, since ambient light reflects from the shiny internal parts of the display and makes viewing of the LED chips difficult. To counter this problem most manufacturers offer their displays in either clear or transparent red packages. The latter approach is for displays operated without an external filter.

Displays with a clear package must be supplemented with an external filter if they are to be viewed in the presence of ambient lighting. There are very few applications where a display will always be in subdued light, so most designs incorporate filters. Filter selection is generally limited to clear red bezels and circularly polarized screens. For very bright ambient light, a combination of colored and polarized filter might prove desirable. As shown in Fig. 4-16, the use of a filter enhances viewing contrast by significantly reducing reflections of ambient light within the LED display.

Before going further, it should be noted that manufacturers have incorporated several manufacturing steps that significantly improve con-

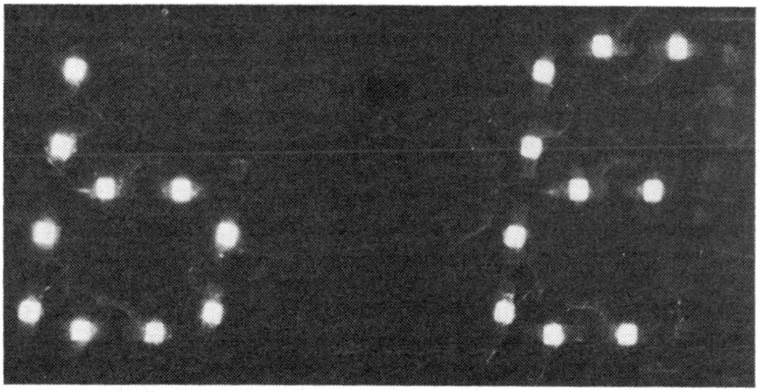

Fig. 4-16. Section of display in Fig. 4-15 with contrast filter added.

trast. The simplest is blacking over reflective metal connections to prevent reflections. More interesting are module-like displays that have small slits for viewing the LEDs. Viewing angle is unaffected, since the LEDs are very close to the slits, but all nonactive components are effectively blocked from view.

The very small size (about 0.1 inch) of many monolithic displays has prompted the incorporation of "bubble lenses" over individual displays or even segments of a multidisplay readout. Sometimes a cylinder lens is employed. Both techniques are designed to increase the apparent size of the display by means of the magnification of the lens. But because of the high f/number required to enlarge the characters, distortion may occur and viewing angle may be affected. Nevertheless, integral expanding optics can prove quite valuable in increasing apparent display size. Fig. 4-17 shows three Hewlett-Packard displays with an integral plastic lens (HP model 5082-7405).

As with displays that incorporate integral filters and masking techniques to improve contrast, many firms offer modules containing several individual displays with integral bubble lenses. The appearance of 5×7 dot arrays is enhanced by means of cast arrays of small bubble

Courtesy Hewlett-Packard

Fig. 4-17. HP 5082-7405 monolithic LED displays with integral bubble lenses.

lenses. Epoxy is frequently employed for making the lenses, and it is possible to cast a lens or lenses over an array as an integral unit in one processing step.

COMMERCIAL DISPLAYS

A great variety of commercial displays is available from LED companies in the United States, England, Japan, and other countries. The mass production of displays has created a large surplus market, and experimenters and hobbyists can obtain both displays and decoding logic at very economical prices from discount companies that advertise in electronic hobbyist publications.

Because of the very large variety of displays offered now and the continual influx of newer units, it is not feasible to provide a complete listing of commercial displays in a general book on LEDs.

Many companies offer extensive varieties of character size, monolithic and discrete construction, self-contained optics, clear or red-tinted potting, and various pin configurations. Several companies offer displays with either integral or modular external logic. These displays can greatly simplify equipment design and speed up assembly.

CUSTOM LED DISPLAYS

For the right price most manufacturers will be willing to custom design an LED display. But if this is not practical, the enterprising engineer or experimenter can design and assemble his own custom display by means of an etched circuit board and discrete red, yellow, or green emitters.

The disadvantage to the custom approach is in fabricating the various LEDs into a usable device, but an important advantage is flexi-

Fig. 4-18. Digitec 268 digital millivoltmeter.

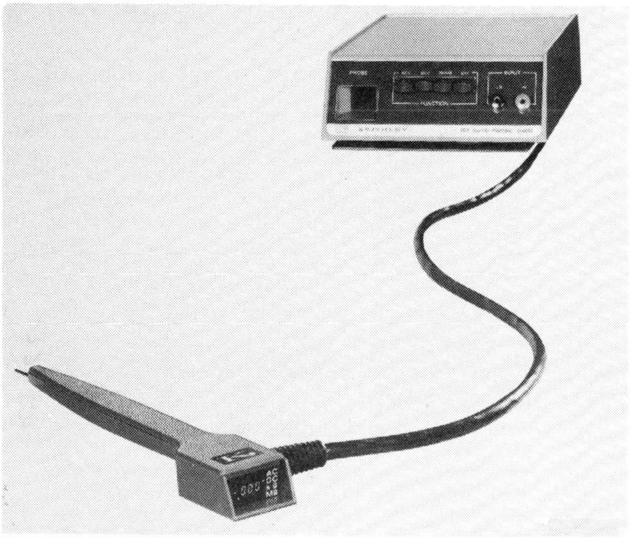

Fig. 4-19. Keithley digital meter with LED readout probe.

bility of design and the possibility of very large character size. Using small diffused-lens emitters a display with a character height of more than an inch and a half can be made, a giant when compared to conventional hybrid arrays. If clear-lens diodes are employed, a red filter should be placed over the array.

APPLICATIONS

Many kinds of electronic equipment now employ LED readouts. The biggest consumer market for displays is miniature calculators. Another consumer application is for electronic clocks.

Electronic test equipment is a fast-growing market for LED displays. A dc millivoltmeter with a clear, easy-to-read LED display is

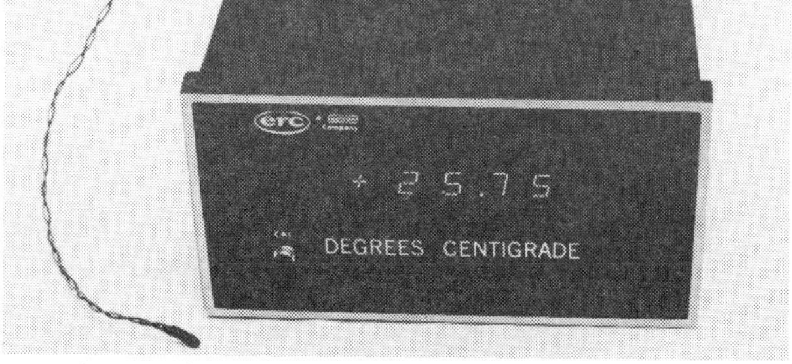

Fig. 4-20. Electronic Research Company digital thermometer.

shown in Fig. 4-18. The meter is made by United Systems Corporation.

Since LED displays are very compact, several instrument companies have considered designing miniaturized test equipment in a probe-like package with built-in readout. While this kind of sophisticated test equipment is not yet available, the next best thing is the readout-in-a-probe concept several companies sell now. One example is of a digital multimeter shown in Fig. 4-19. The self-contained readout simplifies troubleshooting and design work by speeding up use of the instrument.

Test equipment is not the only application for digital displays. There are numerous other uses, as shown by the digital thermometer in Fig. 4-20. Calculators will probably remain the biggest application, but numerous others will be found as prices continue to drop. Ultimately LED displays for radio tuners, inexpensive clocks, automobile dashboards, and other common applications will become a reality.

CHAPTER 5

Detectors, Receivers, and Viewing Devices

The most common detector for visible-light emitting diodes is the human eye, but many applications for visible, and most for infrared, emitters require an electronic detector to sense the presence of the optical beam. This chapter will describe many simple detectors, as well as several specialized units designed for high-sensitivity applications.

In addition to covering electronic detectors that convert the LED signal into an electrical current, this chapter will discuss in detail several ways to make visible an invisible infrared beam from a GaAs or GaAs:Si LED. Viewing techniques can be of great value in using infrared emitters in practical applications.

PHOTORESISTIVE DETECTORS

A wide variety of inexpensive photoresistive cells capable of responding to visible and near infrared radiation is available. These cells are characterized by a resistance that is generally very high in darkness. An incident light beam will cause the resistance of the cell to drop, with the degree of change being a function of the optical flux. For light with a wavelength of from about 500 to 620 nm, cadmium sulfide (CdS) is an ideal photoresistive material. For wavelengths in the red portion of the spectrum, cadmium selenide (CdSe) is best.

Applying the spectral sensitivity of CdS and CdSe to available LEDs, we find that the former material is well suited for green, yellow, and orange emitters, while the latter is ideal for red. In fact, the peak sensitivity of CdSe can be tailored to match the peak emission of red GaAsP and GaP LEDs. CdSe can also be used for detecting near-IR LEDs, but the spectral sensitivity beyond 800 nm is generally below 20 percent. The sensitivity of both CdS and CdSe is summarized in the graph in Fig. 5-1.

Photoresistive cells can be very sensitive. The ratio of dark to light (off to on) resistance can vary from 100:1 to more than 10,000:1.

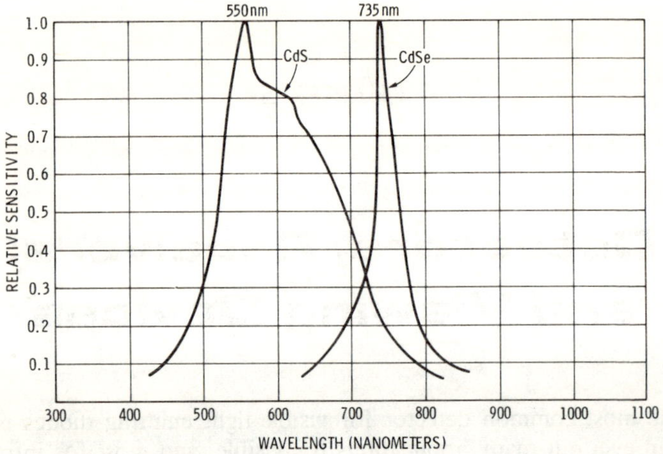

Fig. 5-1. Typical spectral response of CdS and CdSe photoresistors.

But they are very slow and have nonlinear response characteristics when the incident light is increased above a few footcandles. CdSe is usually faster than CdS, but, depending on the light intensity, both materials have response times measured in milliseconds or more. Response times are made faster by increasing the illumination intensity.

An important application for photoresistive cells is in opto-electronic couplers wherein a visible LED supplies an optical turn-on signal for a CdS detector. The purpose of the coupler is to transfer a signal from one point to another without electrical contact, and the opto-coupler accomplishes the job nicely. The turn-on signal need not be a binary "on-off." Since a CdS cell can be linear over a certain range of applied illumination, the opto-coupler can be used to transfer an analog signal.

Like most other detectors, photoresistive cells are not immune to ambient light. When applying the cells in practical applications, it is frequently necessary to provide a mask, baffle, or other device to block stray light. The cells frequently suffer from fatigue or light-history effects. A light level of a specific intensity may affect the photoresistive characteristics of a cell in such a manner that the response to a subsequent level will not necessarily be normal. Light-history effects may take as long as a few seconds to disappear, and sometimes cells can be fatigued for even longer periods.

PHOTODIODES

Pn junction photodiodes comprise the most important variety of sensors available to the LED user. The family includes both *photo-*

voltaic devices such as solar cells and *photoconductive* sensors such as PIN detectors. A specialized photoconductive detector is the *avalanche photodiode.* Each of these classes of pn junction photodiodes is so important to the LED user that they will be described in separate sections. First, just how does a photodiode operate?

When practically any semiconductor junction is illuminated, a small current will be observed. The source of illumination can be either visible light or infrared, but the most common semiconductors, silicon and germanium, have peak responses in the near infrared. The light-sensitivity characteristics of pn junctions have been known since the last century, and in the early 1950s plastic-encased transistors were found to be exceptionally light sensitive. Even common glass-encased silicon diodes are to some degree sensitive to light. Light emitting diodes are no exception to this rule. A fascinating discussion on using LEDs as LED *detectors* will be found toward the end of this chapter.

When light strikes a bulk semiconductor such as CdS or CdSe, photons impinging upon a pn junction create hole-electron pairs. The pairs can be generated on each side of the junction. The energy contained within the illuminating photons causes the electrons to be displaced upward into the conduction band while the holes remain behind in the ground state (valence band). If the two sides of the junction are connected to one another through an external load, electrons will resume their equilibrium by combining with holes on the other side of the junction. The result is a current flow as both electrons and holes move toward the junction. The circuit is completed by the external load.

Thus far we have described a *photovoltaic* detector, one in which a current flow is generated by a pn junction without the assistance of any external power source other than light. However, substantially more current can be made to flow by connecting the junction to an external voltage. If the junction is connected in the normal or forward direction, the current increase will not be great. But connecting the junction in the backward direction will cause a substantial current flow when light is allowed to illuminate it. This technique is called *reverse biasing,* and it is frequently employed with silicon photodiodes. Since the photodiode reverse resistance *decreases* with applied light, it is said to be operated in the photoconductive mode when reverse biased.

In one respect the pn junction photoconductive detector closely resembles the bulk semiconductor photoresistive devices. However, an important difference is that under a constant source of illumination the former device acts as a constant-current source no matter what the bias voltage is. Bulk-effect sensors pass a current that varies with bias voltage because they are actually light-controlled variable resistors.

Photovoltaic and photoconductive junction detectors each have relative advantages and disadvantages. Following descriptions of the most common photodiodes, the merits of the various devices will be described.

SOLAR CELLS

A photovoltaic detector has the ability to generate a voltage in the presence of light, and the most common photovoltaic detector is the silicon solar cell. Solar cells have a peak spectral response in the near infrared at about 850 nm. At the wavelength of GaAs and GaAs:Si LEDs, spectral sensitivity is about 84 percent and 76 percent respectively.

A typical solar cell is shown in Fig. 5-2. Because of their low cost and availability, solar cells are widely used in near infrared LED detection applications. Their rise time is much faster than photoresistive units, with a typical value being 20 microseconds. The size of the cell can affect rise time because the capacitance of the cell increases with area. The relatively large area of solar cells (from 0.2 cm × 0.4 cm to 2.0 cm × 2.0 cm) makes them ideal for collecting LED power without the use of an external lens.

Solar cells can be used in optical communicators, intrusion alarms, optical counting systems, door openers, and other common applications for LEDs. When there is any separation between the solar cell and LED, a transistor amplifier can be used to beef up the signal, but at very close ranges an amplifier may not be necessary for some applications. For example, a 5-mW GaAs:Si LED will generate a current of a few milliamperes in a nearby solar cell.

A valuable application for solar cells in the manufacture and evaluation of LEDs is measurement of LED power output. Many firms which manufacture LEDs employ the solar cell technique to rate both visible and infrared LEDs. If the emission pattern and wavelength of

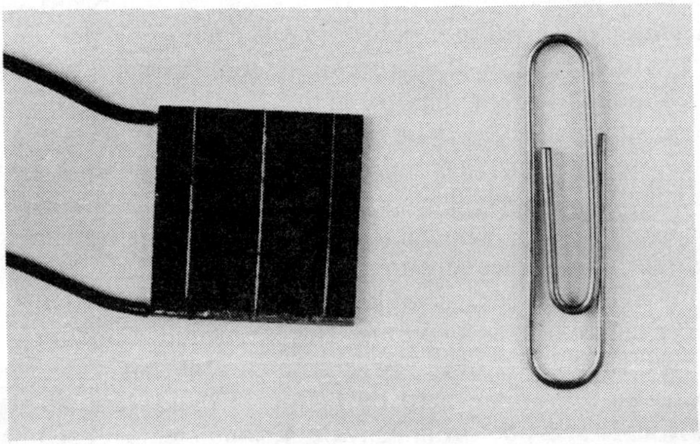

Fig. 5-2. Standard 2 cm × 2 cm silicon solar cell.

a visible emitter are known, the radiometric power of the device can be used to determine various photometric quantities.

PIN PHOTODIODES

The most popular pn junction photoconductive diodes are those which employ several varieties of PIN junctions, and the two most common PIN structures are the Schottky barrier and planar diffused types. Both structures consist of heavily doped p and n regions separated by an undoped region of bulk (intrinsic) semiconductor. The resistivity of the doped material can be very small and is typically less than 0.1 ohm-cm. However, the intrinsic region may have a resistivity of from 10 ohm-cm to 100,000 ohm-cm.

When an external bias is connected to a PIN photodiode in the reverse direction, an electric field will be generated within the bulk silicon intrinsic region. With no light present, a current (the *dark current*) of only a microamp or so will flow; however, as external illumination is applied, the current will be increased substantially. A typical PIN photodiode will pass a current of from 0.4 to 0.6 microampere per microwatt of applied illumination at the peak response wavelength of the diode (about 900 nanometers). This figure of merit is called *responsivity.*

PIN diodes can be fabricated in a wide variety of configurations. For example, the three large Harshaw detectors in Fig. 5-3 actually incorporate four separate detectors on a single substrate. These *quadrant detectors,* as they are called, are very valuable for sensing the position of an incoming spot of light. For sensing light intensity along

Courtesy Harshaw Co.

Fig. 5-3. Standard and quadrant photodiodes.

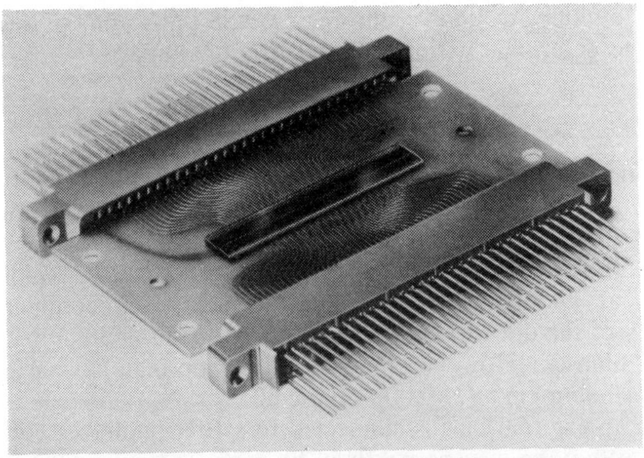

Courtesy United Detector Technology

Fig. 5-4. Linear photodiode array.

a line, a detector such as the one shown in Fig. 5-4 can be employed. And for specialized applications requiring a mosaic of sensitive detectors, the 100-element array in Fig. 5-5 can be used.

SCHOTTKY BARRIER PIN PHOTODIODES

A Schottky barrier PIN photodiode is shown in Fig. 5-6. The device is made by applying an aluminum layer to one side of a slab of intrinsic silicon and a transparent gold film to the other side. Both metal

Fig. 5-5. United Detector Technology 100-element photodiode array.

Fig. 5-6. Schottky barrier PIN photodiode.

layers serve as contact points, but the parameters of the gold one are critical, since the light to be detected must pass through it. The gold layer is usually an evaporated film only about 150 angstroms thick. The thickness of the silicon wafer can vary from about 0.125 to 1.0 millimeter.

For the Schottky barrier diode to operate in an optimum manner, an electric charge should extend across the intrinsic bulk silicon to both p and n regions at the contacts. When this occurs the diode is said to be fully depleted. The thickness of the intrinsic region is directly related to the voltage required for depletion. To calculate the penetration (depletion depth) of the field, a simple formula can be used:

$$d = \frac{\sqrt{RE}}{2} \qquad (5\text{-}1)$$

where,

d is the depletion depth in microns,
R is the silicon's resistivity,
E is the applied reverse bias voltage.

This formula can be valuable for calculating required voltage for full depletion when both resistivity and the intrinsic region thickness are known.

DIFFUSED PIN PHOTODIODES

The structure of a diffused PIN photodiode is probably more familiar to most readers. The back contact side of both kinds of diodes consists of an evaporated aluminum layer. However, the diffused diode incorporates conventional masking and diffusing techniques for the front side. First a layer of silicon dioxide is evaporated over the intrinsic silicon wafer. Then a hole is etched away to reveal some of the bulk silicon. Next, an impurity such as boron is diffused into the bulk material to form a junction. Contact is made directly to the high-conductivity boron diffused region by means of one or more bonded wire electrodes. The diodes in Fig. 5-3 are diffused PIN units as is the unit in Fig. 5-7.

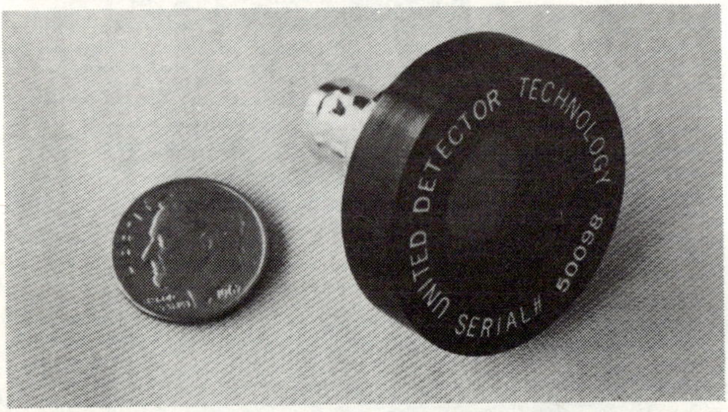

Fig. 5-7. United Detector Technology diffused PIN photodiode.

GUARD RINGS

Both Schottky barrier and diffused PIN photodiodes exhibit a dark current, and since the ultimate sensitivity of a photodiode is dependent on a high signal-to-noise (SNR) ratio, it is important to reduce current leakage to an absolute minimum. Part of the dark current is due to bulk leakage, but another part is caused by surface leakage over the sides of the diode and across the junction. Surface leakage, sometimes called channel leakage, can be eliminated by means of special fabrication techniques or with an additional electrode called the guard ring. The guard ring is fabricated at the same time as the active-area junction. On a diffused PIN photodiode, a circle is etched in the silicon dioxide layer around the active region and both are diffused simultaneously. In fact, both regions become pn junctions. The guard ring is activated by biasing it at the same voltage as the active area. This causes any surface leakage to flow through the guard ring and not the active area. Since bulk leakage is as much as 100 times less than the surface leakage, the guard ring can significantly improve the performance of a PIN diode by greatly increasing its SNR.

SCHOTTKY BARRIER vs DIFFUSED PHOTODIODES

As might be expected, both Schottky barrier and diffused PIN photodiodes have relative merits and demerits. The former devices are more temperature sensitive and have less immunity to high light levels than is true of diffused diodes. The difference in maximum acceptable light intensity levels is due to the possibility of damage to the gold film electrode over the active area of Schottky barrier devices. On the other side, the heat required to form diffused diodes can alter the characteristics of the high-resistivity-silicon intrinsic region. Also, Schottky barrier diodes can be constructed in much larger sizes than their diffused counterparts.

Both types of diodes have very low noise, and both have response times measured in nanoseconds. A very important characteristic possessed by these types of diodes is their high degree of linearity, and both Schottky barrier and diffused diodes are linear over about seven decades of light intensity. This feature is particularly important in applications where light levels will vary greatly. It is also very important in optical receiver systems designed to be immune from effects of sunlight without the need for an external filter.

PHOTOVOLTAIC vs PHOTOCONDUCTIVE OPERATION

Most kinds of silicon photodiodes can be operated in either the photovoltaic or photoconductive mode. The former operating condition is characterized by the absence of dark current and is therefore well suited to the detection of very low light levels. In situations where

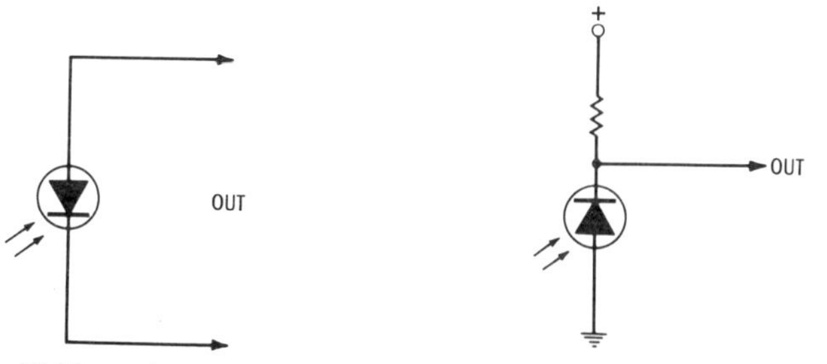

(A) Photovoltaic mode. (B) Photoconductive mode.

Fig. 5-8. Photovoltaic and photoconductive operation.

the light is pulsed, the photoconductive mode is superior. The dark current, a dc parameter, can then be ignored by capacitively coupling the detector to an amplifier or circuit. The reverse biased photoconductive mode will give twice the responsivity and far faster response times than the photovoltaic mode. Summing up, choose photovoltaic operation for very low level dc measurements and photoconductive for high-frequency pulsed measurements.

The circuit configurations for photovoltaic and photoconductive operation are shown in Fig. 5-8. The load resistor in the photoconductive mode should be small for high-frequency response and large for maximum sensitivity.

AVALANCHE PHOTODIODES

If the reverse bias on a diffused PIN photodiode is increased beyond a certain point, the diode may break down and conduct without the presence of external light. This phenomenon is called *avalanching,* and if the diode is operated within a few millivolts of breakdown, an incoming optical pulse can trigger the avalanche. Since the required signal can be very small, avalanche detectors have high sensitivity.

Conventional PIN diodes are not operated in the avalanche mode, as the thickness of the intrinsic region can be such that a very high voltage may be required to achieve the avalanche condition. Special structures are adopted in order to permit avalanche operation without damage to the device. General Electric, for example, employs a contoured-surface diode that has an avalanche breakdown of from 1800 to 2600 volts.

The most important characteristic of the avalanche photodiode is internal gain. Whereas a conventional photodiode responds with a current of perhaps 0.5 ampere per watt of incident radiation, avalanche detectors may have a typical response of 100 and up to 275 amperes per watt at the peak radiation sensitivity of 900 nanometers. The reason for this high gain is the avalanche process itself. The arrival of but a few photons at the sensitive surface of an avalanche detector can cause the recombination of a few hundred electron-hole pairs. The result is *avalanche multiplication.*

For all its impressive abilities, the avalanche detector has an imposing drawback: operating voltage must be carefully maintained to keep the device just below the breakdown point. Since the avalanche point varies with temperature, voltage regulation is a significant problem. In a non-temperature-regulated power supply, a shift of just 1 or 2 degrees Celsius can take the detector out of the avalanche region.

Texas Instruments and General Electric both solve the critical voltage- and temperature-regulation problem of avalanche photodiodes by manufacturing a variety of power supply and amplifier modules. The modules provide temperature-regulated bias voltage and, in some cases, amplification. Some of the Texas Instruments modules, which are quite compact in physical size, even include an integral thermoelectric cooling element for the photodiode.

Avalanche detectors offer such high sensitivity and gain that one major manufacturer reports a trend away from injection laser diodes and conventional PIN photodiodes to high-power LEDs and avalanche-detector modules in optical radar and communication systems. The avalanche-detector system offers such a significant increase in sensitivity that the high power output of the laser is just not needed. Using the LED instead of the laser means simpler power-supply requirements, lower current consumption, less component degradation, and lower cost. For the ultimate in sensitivity, of course, an injection laser can be used instead of an LED.

PHOTOTRANSISTORS

As noted earlier in this chapter, in the early 1950s it was found that plastic-encapsulated transistors could be exceptionally sensitive to light, and even with no base current such a transistor might exhibit current flow in the presence of light. For quite a few years engineers and experimenters exploited this unique characteristic of transistors in a variety of light-controlled circuits. Since few commercial phototransistors were available, it was a common practice to simply remove the protective case from a common transistor and employ it as a light sensor. Frequently a number of units were tried until one was found that had suitable sensing characteristics.

For the past several years, a great variety of phototransistors have been available commercially at very low cost. These devices are conventional bipolar and field-effect transistors fabricated to expose a region free of metal contacts and other light-blocking structures. Most early phototransistors were housed in conventional metal transistor cans with a flat glass window or lens in one end, and some units were housed in miniature glass tubes. More recently all-plastic packaging has become very popular because of low cost and ready availability, but for applications where mechanical rigidity is important, hermetically sealed metal can units are employed.

Operation of a conventional bipolar phototransistor is dependent on the photoeffect at a pn junction. If an exposed pn junction is made the base-collector junction of a transistor, any light-induced current will become the base current of the transistor. Since no external current is required for transistor action to occur, there is no requirement to apply external base-bias voltage, and many phototransistors are made with only collector and emitter leaders. The characteristics of the phototransistor can be modified by application of external base voltage, however, and some applications require the additional flexibility of an external base connection.

The chief advantage of a phototransistor over a photodiode is gain. The current gain of a transistor will hold, whether the base current is supplied by an external source of electrons or of protons, so phototransistors can be very sensitive devices. Unfortunately, transistor gain is not linear and varies with collector current, and the result is a nonlinear response to incident light. For very small light levels the gain will be relatively linear, but saturation rapidly occurs as incident light passes a certain intensity. As a result, phototransistors cannot be used to detect modulated light beams in the presence of high ambient light levels unless an optical filter is used to eliminate the interfering source.

Another disadvantage of phototransistors is relatively slow response times. Because of the internal capacitance of the junctions, phototransistor response times are measured in microseconds. As with pn photodiodes, response time is largely dependent on the value of the load resistor, since junction capacitance must be charged or discharged through the load. By using a very small load resistor, response speed can be increased; however, this affects the characteristics of the transistor, since maximum sensitivity occurs with large values of load resistance.

One way to improve the response of a phototransistor is to couple the load resistor to it by means of an impedance converter. In this way a large value of resistance can be used for a load, while preserving the low impedance necessary for the transistor to exhibit fast response time. A suggested circuit is shown in Fig. 5-9. The parts values shown in the circuit diagram are suggested values, and many other components can be substituted with no problem. For maximum frequency response the load resistor should have a value of about 10,000 ohms. More sensitivity can be obtained with a larger value of resistance, but an increase beyond about 1,000,000 ohms will begin to cause a significant reduction in frequency response.

Because of their nonlinear response characteristics, phototransistors are more sensitive when a slight base-bias current is applied. Since most of these devices have no base lead, it is not possible to apply electrical base bias, and an optical source must be used. One way is to simply place a small LED near the transistor and adjust the LED current until optimum transistor sensitivity occurs. This optical biasing technique may not be necessary if a small amount of ambient light is available.

DARLINGTON PHOTOTRANSISTORS

For even higher sensitivity than possible with a phototransistor alone a Darlington phototransistor can be employed. Several photo-

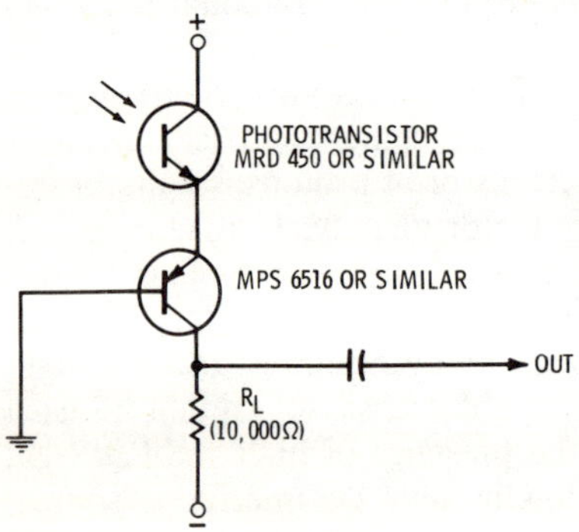

Fig. 5-9. Phototransistor speed-up circuit.

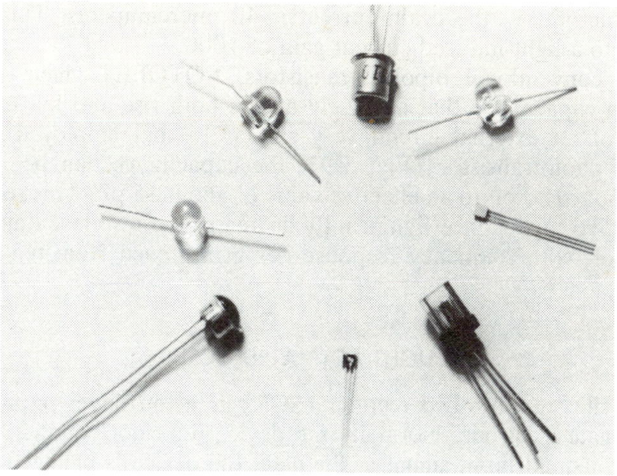

Fig. 5-10. Assortment of low-cost transistors.

transistors and Darlington phototransistors are shown in Fig. 5-10. Conventional phototransistors have a sensitivity of typically 0.1 to 0.8 mA/mW/cm². With the Darlington configuration, these values can be increased to as much as 4.0 mA/mW/cm².

LIGHT-SENSITIVE FETs

The advantages of the field-effect transistor (FET) can be preserved in a light-sensing mode by simply providing a glass window and an open junction region. The result is the "PhotoFET." Teledyne Crystalonics manufactures a line of these devices under the trade name FOTOFETS. Their chief advantage is very high input impedance and fast response time, and in the proper circuit arrangement a typical FOTOFET will exhibit a response time of 30 nanoseconds. This gives the device a response time falling between PIN diodes and conventional phototransistors, while preserving the gain feature of a transistor. Furthermore, while the gain of a phototransistor may be 100 or 200, a typical FOTOFET may have a gain of several million. Another advantage of the FOTOFET is a more linear response to incoming light. While phototransistors quickly become nonlinear and saturate at high light levels, FOTOFETs maintain a relatively high degree of linearity over a large range of optical-intensity fluctuations.

The high gain of a FOTOFET is illustrated by this example: if a Teledyne Crystalonics FF102 FOTOFET is illuminated with one foot-candle, a gate current of 40 nanoamperes will be caused to flow through a one-megohm gate resistor. This corresponds to a voltage of 40 millivolts. Since the FOTOFET has a transconductance of about

1000 micromhos, the drain current is 40 microamperes. This corresponds to a light-induced current gain of 1000.

Like conventional bipolar transistors, FOTOFETs have internal junction capacitance that adversely affects both rise and fall time. By means of an external circuit very similar to that employed for the bipolar phototransistor (Fig. 5-9), the capacitance can be reduced from about 180 pF to an effective value of about 15 pF. This reduction in effective capacitance significantly decreases the response time of the transistor, with frequency response being improved from ten to fifty times.

LIGHT-ACTIVATED SCRs

The silicon controlled rectifier (SCR) is a four-layer pnpn device with a gate electrode. Normally the device does not conduct, but by placing a small input signal at the gate, the device is changed from a nonconducting state to a low-resistance conducting state. This switching action can be very valuable in numerous applications, and the SCR can frequently be employed as a direct replacement for the electromechanical relay. Switching time is much faster than electromechanical devices, and the inherent advantages of a solid-state device (compact size, long life, mechanical rigidity, etc.) are preserved.

It so happens that a beam of light can be made to trigger an SCR just as if a gate voltage had been applied. Several companies have exploited this characteristic with a variety of light-sensitive SCRs. While light-activated SCRs (LASCRs) do not retain all the characteristics of conventional pnpn switches, they are valuable in several interesting and practical applications. Response time may be as high as 30 microseconds, but this can be reduced below a microsecond by employing a high-intensity pulse of light. The light intensity required to trigger the device can be varied by simply changing the gate current.

DETECTOR-AMPLIFIER MODULES

In the discussion on avalanche photodiodes it was noted that some manufacturers offer complete subsystems containing detector, regulated power supply, and amplifier as a modular package. The same is true for phototransistors and PIN photodiodes, with half a dozen firms manufacturing miniature detector-amplifier modules. One such device is the RCA CA3062 monolithic silicon integrated circuit containing a pair of Darlington phototransistors and a switching amplifier. The entire unit is housed in a glass-windowed TO-5 package, and only a few external components are required to properly operate it.

RCA has designed the CA3062 to be compatible with its line of infrared-emitting diodes. The unit has a typical switching speed of 163 microseconds (turn-on time plus rise time). Total turn-off time plus fall time is 63 microseconds. The device is primarily intended for ON-OFF applications, but by means of proper circuit connections the unit will operate as a linear amplifier of incoming light signals. Full flexibility is permitted since the Darlington pair is connected to the amplifier by means of the IC's external pins. The amplifier and photodetector section occupy the same silicon subtrate but are not connected to one another internally.

For faster response times, PIN photodiodes plus amplifier modules can be employed. The tradeoff is cost. While the CA3062 can be purchased for several dollars, PIN detector-amplifier modules may cost $100 or more. Many of the PIN diode manufacturers offer detectors with integral amplifiers. The United Detector Technology UDT-400 and UDT-500 contain an integral high-speed photodiode and operational amplifier. The UDT-500 employs a large-area photodiode and a FET input stage to the operational amplifier. In a typical application, a one-microwatt signal from a GaAs LED will provide an output voltage of 0.4 volt when a one-megohm load resistor is employed. The UDT 500 and a test box are shown in Fig. 5-11. The company also makes the UDT-600, a unit optimized for high bandwidth laser and LED detection applications.

The advantage of an operational amplifier in the PIN diode module is linearity of output. By simply varying the feedback resistor in the amplifier, linearity can be preserved over seven decades of input light intensity. This permits a PIN diode/operational amplifier module to

Fig. 5-11. United Detector Technology UDT-500 photodiode-amplifier module and UDT-505 test box.

be employed as a power meter by simply connecting a meter to the output and a number of known resistors to a feedback-resistor switch.

EG&G and MERET, Inc., also make photodiode modules. EG&G's HAD-1000 incorporates a PIN photodiode (SGD-100A) and an operational amplifier with a FET input. While the HAD-1000 and the UDT-400 and 600 are housed in TO-5 style packages, MERET offers integral photodiode amplifiers in both TO-5 and flat-pack configurations. The latter device, the FDA425, measures 9.4 mm (0.37 inch) square and is only 0.76 mm (0.075 inch) thick. The manufacturer claims a detection capability of 70 nanowatts with a SNR of 10:1 at 905 nanometers from a GaAs LED. The device requires 9 volts at 5 mA for proper operation. Two MERET detector modules are shown in Fig. 5-12.

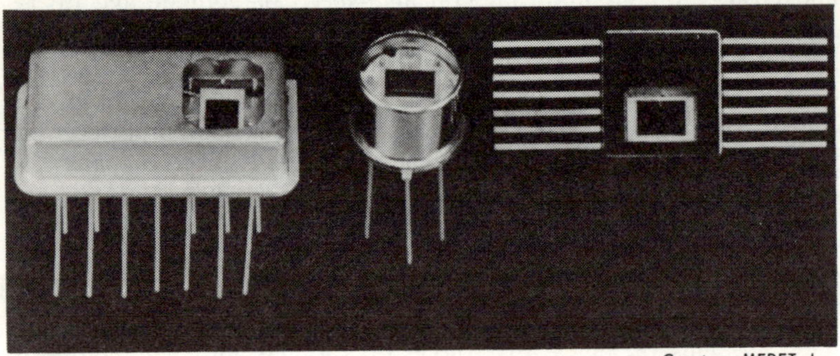

Courtesy MERET, Inc.

Fig. 5-12. MERET, Inc., photodiode-amplifier modules.

LEDs AS DETECTORS

As was noted earlier, semiconductor junctions sometimes are capable of both generating and detecting light. The LED is no exception, and it is possible to employ an LED to detect the output from a similar LED. For best results, both LEDs should have the same bandgap, a requirement easily solved by using devices with identical semiconductor material.

Fig. 5-13 shows circuits for testing LEDs as both photovoltaic and photoconductive detectors. The LED employed as an optical source is connected to a power supply and biased with an appropriate current. The LED used as a photovoltaic detector is connected to a voltmeter. In a typical test, a GE SSL-55C caused an identical LED to generate a 0.7 volt signal across a high-impedance voltmeter. A silicon solar cell with the same surface area as the lens of the LED generated only about 0.05 volt for the same conditions.

The author has measured the quantum efficiency of an RCA SG1004 in the detector mode by using a calibrated SG1004 LED, emitting 3.0

mW. A second SG1004 was connected to a current meter, and the two diodes were placed directly adjacent to one another. The resultant quantum efficiency for the LED detector was 34 percent.

Silicon PIN diodes may have quantum efficiencies of 80 percent or more, but the LED test was impressive, since the optical coupling between the two LED chips was not ideal. In fact, much of the radiation probably never reached the detector chip.

The significance of using an LED as a detector is potentially great. For example, it might be possible to fabricate optical logic devices where the active devices function as both detectors and transmitters. Another important application would be in optical communications, since a single LED could serve as both a source and a detector of radiant energy. The significance is the saving in parts and the need for only one lens. The latter advantage can be of significant importance in miniature optical communication systems where space is at a premium and in systems employing fiber optics.

To determine whether or not an LED can be used in an optical receiver, match up LED detector sources with one another using the arrangements in Fig. 5-13. Note that while visible emitters also work, best results are had with GaAs:Si units. Then determine the efficiency of the LED as a working detector by trying it in both the photovoltaic and photoconductive modes. In one application tried by the author a pair of GE SSL-55Cs were used in the photoconductive mode. With one LED supplied with 100-mA current and delivering 6 mW of near infrared, the other was reverse biased with 3.0 volts and connected to a one-megohm load resistor. With the LEDs separated 5 cm (about 2 inches), the "detector" LED generated a current of 2.4 microamperes. While this is not as high as would be desired for some roles, a silicon PIN diode in the same optical system gave a current only slightly higher (2.8 microamperes). The surface area of the PIN photodiode was larger than that of the LED, and imperfect focusing of the self-contained lens of the LED might explain some of the difference.

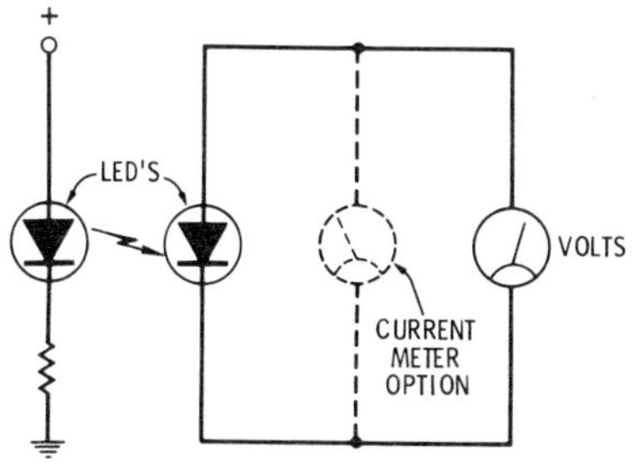

(A) Photovoltaic mode.

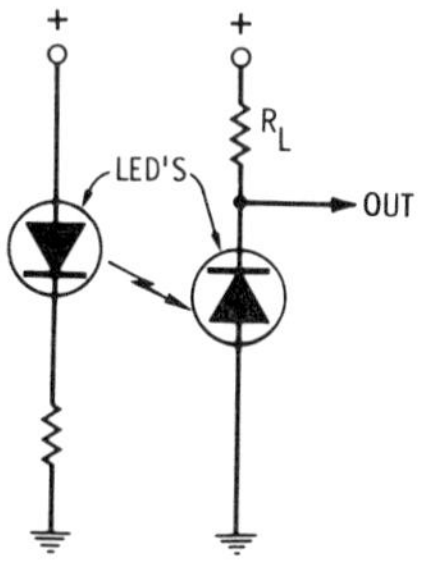

(B) Photoconductive mode.

Fig. 5-13. Test circuits for LEDs operated as detectors.

VACUUM DETECTORS

Vacuum phototubes and photomultipliers such as the one shown in Fig. 5-14 are still employed in some detection applications. Photomultipliers have very high sensitivity because of internal electron multiplication, but they are bulky and fragile, they require a high operation voltage, and they must be shielded from ambient light. Photomultipliers are also generally more expensive than their solid-state counterparts.

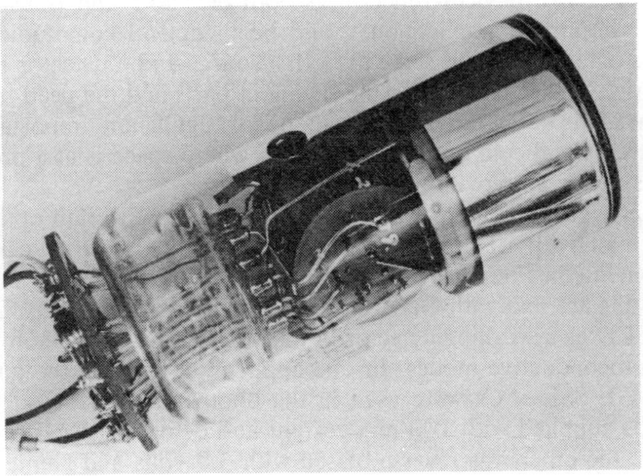

Fig. 5-14. Photomultiplier tube.

For applications where ultrahigh sensitivity is required, the ability to detect single photons makes the photomultiplier hard to beat. By means of suitable amplification, however, some PIN diode configurations (both conventional and avalanche) can approach tube sensitivity. Table 5-1 compares some of the characteristics of solid-state detectors and photomultiplier tubes.

CALIBRATED DETECTORS

A large-area detector can be of great value when measuring the power output of an LED or other optical source. Power output information is important for calculating the transmitting range of optical communications systems, the detection range of optical ranging systems, and optimizing various LED-detector geometric arrangements. Several companies market standard PIN photodiodes that have been calibrated using generally accepted techniques, but their cost may be $100 or more. If a calibrated thermopile is available, it is possible to accurately calibrate an ordinary solar cell and save the expense of a

Table 5-1. Typical Detector Characteristics

Detector	Peak Spectral Response (nanometers)	Sensitivity (mA/mW)	Response Time (microseconds)	Dark Current (nanoamperes)
Photomultiplier	200–900	0.1	0.003	.01
Solar cell	820–900	0.25	20	—
Phototransistor	800	0.8	3	25
Fotofet	950	10.0	0.05	1
Pin photodiode	900	0.5	0.001	2

commercial calibrated detector. Calibrated thermopiles may sometimes be borrowed from research laboratories and universities.

SOLAR-CELL CALIBRATION

Calibration can be a very difficult procedure, and different techniques may produce a variety of calibration values for the same cell. One technique is to employ a bright source of wideband illumination and a series of narrow-bandpass interference filters peaked for each of the wavelengths at which it is desired to evaluate LEDs. The filters are placed, one at a time, between the broadband source and a calibrated thermopile (calorimeter). The thermocouple has the ability to absorb all the incident radiation, convert it to heat, and generate a voltage that is directly related to power but independent of wavelength.

When the filter measurements with the thermopile are completed, the procedure is repeated, with the thermopile replaced by the solar cell. The measurements are then repeated while measuring the short-circuit current of the cell by means of a microvoltmeter and a very small resistor across the cell. The output of the cell is converted to power by multiplying the square of the short-circuit current by the value of the small load resistor. Finally, the cell efficiency at each wavelength is calculated by dividing cell power by the actual power as measured by the calibrated thermopile.

Another calibration technique employs a standard lamp that has a known radiant-power output. This technique can be tricky since the lamp output is dependent on applied voltage, and its power supply must therefore be highly regulated. Both techniques are only as accurate as the calibration of the reference source or detector, of course; thus, it is desirable to employ devices with calibration traceable to the National Bureau of Standards.

As a last resort, still another calibration technique may be employed. Since it is the least accurate of those which will be described, it is the least desirable. However, this technique may give sufficient accuracy for many noncritical applications.

This final calibration technique employs the sun as a "standard lamp." Actually, the sun can be considered as an accurately calibrated standard lamp, particularly since its spectral and power output have been measured by numerous researchers. However, the earth's constantly changing atmosphere and slightly elliptical orbit interfere with power reaching the earth from the sun and cause distortions in any calibration based on sunlight. The fluctuations in energy reaching the earth vary as a function of wavelengths falling in water-absorption bands. GaAs:Si LEDs with a wavelength of 940 nm fall within one of these bands. Therefore, because of variations in atmospheric humidity, calibration will not necessarily be as accurate as would be desired.

INFRARED-VIEWING DEVICES

Because of their high power-output capability, infrared-emitting diodes made of GaAs and GaAs:Si are frequently employed in applications such as optical communications, intrusion alarms, and optical ranging systems. Unfortunately the invisible nature of the infrared beam from these diodes can make them difficult to work with. For example, the optimum placement of a collimating lens can be difficult if the resultant beam cannot be observed.

Several different viewing systems are available that convert invisible infrared into visible light. The least expensive are infrared-sensitive phosphor screens such as those manufactured by Eastman Kodak and Optical Engineering. More costly but more useful are image-converter tubes.

IR PHOSPHOR SCREENS

A relatively economical infrared-viewing device is the IR-sensitive phosphor-coated screen. Eastman Kodak manufactures phosphor-coated screens ranging in size from 2 × 3 inches to 20 × 24 inches. The smallest screen presently costs about $25, and while this may seem high it is much more economical than commercial image-converter tubes. The Kodak phosphor cards must be activated by predominantly blue light (400 to 550 nm) for a brief time before use, and only a few seconds are required if the screen is placed near an ordinary fluorescent tube. Peak sensitivity for the Kodak material is near 1000 nm, with GaAs (900 nm) and GaAs:Si (940 nm) LEDs falling at 60 and 80 percent of spectral sensitivity, respectively. When struck by a near-infrared beam, the phosphor glows orange for from 5 seconds to several minutes, depending on the level of excitation. Thereafter a dark negative image of the infrared pattern that exhausted the phosphor will be seen if the screen is uniformly illuminated by near infrared.

A Kodak phosphor screen was used to make the two photos in Fig. 5-15. In Fig. 5-15A, a GE SSL-55C is shown in normal room light with no forward current. In Fig. 5-15B, forward current has been applied to the LED and the room lights have been subdued. The characteristic central spot and halo of the SSL-55C are clearly visible (see Fig. 2-20).

Kodak publishes two pamphlets describing its IR phosphor screens. Kodak Pamphlet U-70 describes the various screens that are available and provides detailed technical specifications. Pamphlet U-76 provides information on photographing beam patterns on the screens in order to provide a permanent record.

Optical Engineering also manufactures infrared phosphor screens. The company's IR Display Plate consists of an upright aluminum frame coated with three different phosphors sensitized for different infrared wavelengths. Like the Kodak product, the IR Display Plate is activated by exposure to natural or artificial light for a few seconds before use. In tests with GaAs and GaAs:Si LEDs, two of the three phosphor coatings exhibited good response characteristics.

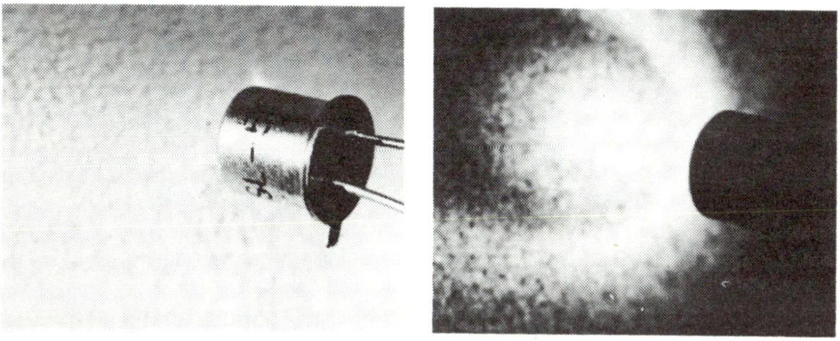

(A) LED off. (B) LED on.

Fig. 5-15. Infrared LED and Kodak IR phosphor screen.

IMAGE-CONVERTER TUBES

From the previous brief discussion on vacuum phototubes, it will be recalled that photoemissive substances have the property of emitting electrons in response to an oncoming beam of light. If one end of an evacuated tube is coated with photoemissive material and a high-voltage electrostatic field is applied across the tube, it is possible to direct the electrons emitted when the photoemissive surface is struck by light back toward the end of the tube. With suitable regulation of the electrostatic field the electron pattern at the rear of the tube will duplicate the light pattern of the light-sensitive face of the tube. These electrons can be used to stimulate a phosphor-coated screen at the rear of the tube.

If the photoemissive surface is sensitive to infrared and if the viewing phosphor converts the energy of impacting electrons into visible light, the image tube described above becomes an image converter. That is, it converts invisible infrared into visible light. Typical image-converter tubes require 15,000 volts dc at low current for proper operation. Some tubes incorporate a central electrode called the *focusing electrode*. Others achieve self-focusing by means of suitably shaped primary electrodes.

Fig. 5-16. IR image-converter tubes and power-supply module.

Several kinds of image-converter tubes are commercially available. Some are shown in Fig. 5-16. Prices are generally above $50 for new tubes, but surplus tubes can be purchased for less than $10. A few companies offer complete infrared-viewing systems like the one shown in Fig. 5-17. These devices are available in a variety of hand-held configurations from FJW Industries and Electrophysics Corporation.

Fig. 5-18 illustrates the resolution of a commercial image-converter tube. In the figure, the author's wife is shown through a converter tube, with a single candle providing the only illumination. The figure was made by photographing the screen of the tube with ordinary panchromatic film.

The experimenter or engineer with a limited budget can assemble an infrared image-converter system for less than about $25. In addition to a converter tube and a high-voltage power supply, lenses of appropriate focal length and f/number will be required for focusing the scene being viewed onto the sensitive face of the tube and for viewing the output.

Fig. 5-17. Commercial IR viewer in use.

Image Intensifiers

Two or more conventional image converters can be cascaded one behind the other to provide an amplification of several hundred or more. In this manner, scenes can be viewed in darkness so long as a small source of indirect illumination (moon or starlight) is available. This high degree of sensitivity has earned infrared image intensifiers the popular namc "Starlight Scopes." Viewing a nighttime scene with

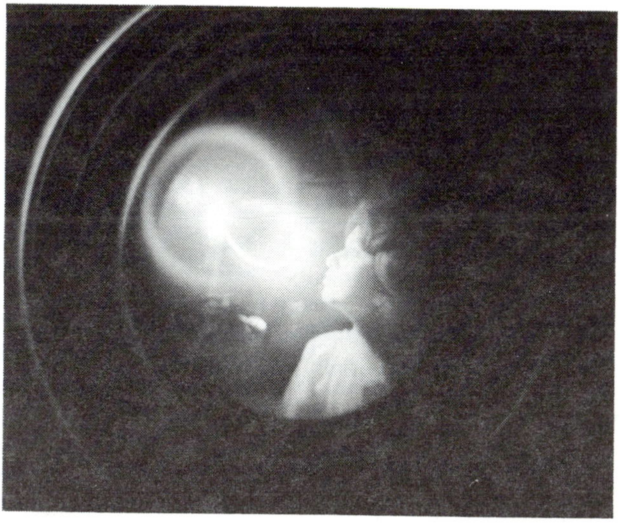

Fig. 5-18. Model photographed through IR viewer with single candle providing illumination.

a starlight scope is an eerie experience. Though the unaided eye sees nothing more than shadows and blackness, the scope provides a greenish image with good resolution of practically everything in its viewing field.

Image intensifier or amplifier systems would be bulky and unwieldy if it were not for fiber optics. Using conventional optics, a complex series of lenses would be required to transfer the image from one tube to another. But by fusing millions of fiber-optic strands into a thin plate to which the rear of one tube and the front of the next can be matched, it is possible to transfer images coherently and directly. These fiber-optic faceplates are quite costly when purchased new, but their large-volume production for military and industrial night-viewing systems has resulted in their availability on the surplus market for a fraction of their original cost.

Image-intensifier systems in portable configurations have found use in military and police applications where the invisible illumination required for their use is supplied by infrared LEDs. GaAlAs, GaAs, and GaAs:Si diodes have been employed with varying degrees of success, but GaAlAs diodes are preferred because their wavelength can be tailored to match the peak spectral sensitivity of the tubes (about 850 nm).

Microchannel Intensifiers

Conventional image-converter tubes suffer from the same disadvantages as most other vacuum tubes: bulky size, fragility, and requirement for a high voltage. Recently a new kind of image-conversion device, the microchannel image intensifier, has received a great deal of

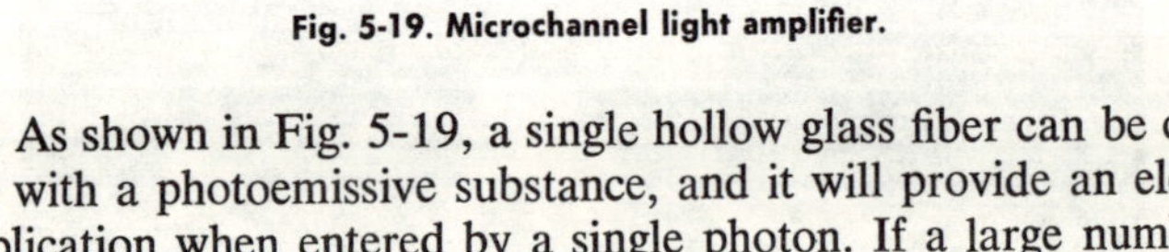

Fig. 5-19. Microchannel light amplifier.

study. As shown in Fig. 5-19, a single hollow glass fiber can be coated inside with a photoemissive substance, and it will provide an electron multiplication when entered by a single photon. If a large number of hollow fibers or capillaries are bonded together into a plate of microchannels, an infrared scene striking one side of the plate can be converted to visible light on the other side. As with conventional image tubes, the viewing surface is a phosphor that emits visible light when struck by electrons.

Microchannel image converters have important advantages over single-stage conventional tubes. The most significant is that the microchannel provides amplification with but a single stage. Additionally, the channel plate device is very compact and light in weight. The one shown in Fig. 5-20, a Bendix model, weighs only a few ounces and is mounted in a robust plastic mount. Their major disadvantage is cost

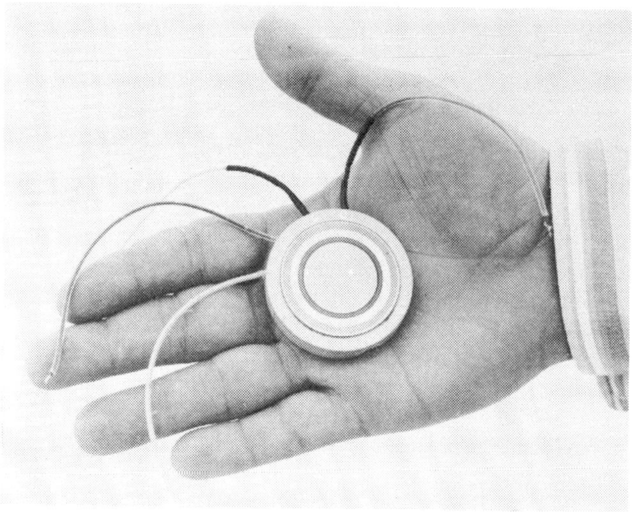

Fig. 5-20. Bendix BX 9025 microchannel image intensifier.

and requirement for a high operating voltage. The voltage requirement is under that of conventional tubes (typically 5000 volts dc), but cost ranges up to a few thousand dollars or more. As manufacturing techniques are improved, costs should come down. Since microchannel plates provide an ideal viewing device for infrared emitters, this will be a welcome development.

Solid-State Image Converters

An even more recent development than the microchannel converter is the self-scanned semiconductor-diode array. A development of Bell Laboratories, this technique has already permitted the demonstration of a completely solid-state television camera by half a dozen companies now manufacturing self-scanned arrays. The present devices must be coupled to a conventional television cathode-ray tube to achieve image conversion, but there is reason to expect that the same technology that produced the light-sensing array may also produce a light-emitting array. This development would be of vast significance, as it would permit all-solid-state tv.

The significance of self-scanned arrays to infrared LEDs is great, particularly since GaAs LEDs closely match the peak spectral sensitivity of the silicon making up the array. If and when solid-state detector-emitter arrays become available, they will be useful in viewing infrared-LED beam patterns as well as for night-vision viewing devices that use infrared illumination. Potentially, cost is low since conventional semiconductor-processing techniques can be used to make the self-scanned arrays. Additionally, the operating voltage is much lower than that required for vacuum image-conversion tubes.

INFRARED PHOTOGRAPHIC FILM

Eastman Kodak manufactures several types of infrared-sensitive film that respond to the radiation from GaAs and GaAs:Si LEDs. While the delay in processing can prove bothersome, the permanent record and high resolution of photographic film can prove invaluable in evaluating infrared-emitting diodes. Several figures in this book are from IR film and provide a good indication of its capabilities.

Two classes of infrared-sensitive film are available. The least expensive is Kodak standard and high-speed infrared film. This material provides a black and white image. For color applications, Kodak infrared Ektachrome can be used. This film includes normal yellow and magenta (reddish purple) layers and a third cyan(blue) layer. The cyan layer is infrared sensitive. Conventional Ektachrome E-4 processing can be used with this film. As shown in Fig. 5-21, both kinds of film are available in standard 35-mm cartridges.

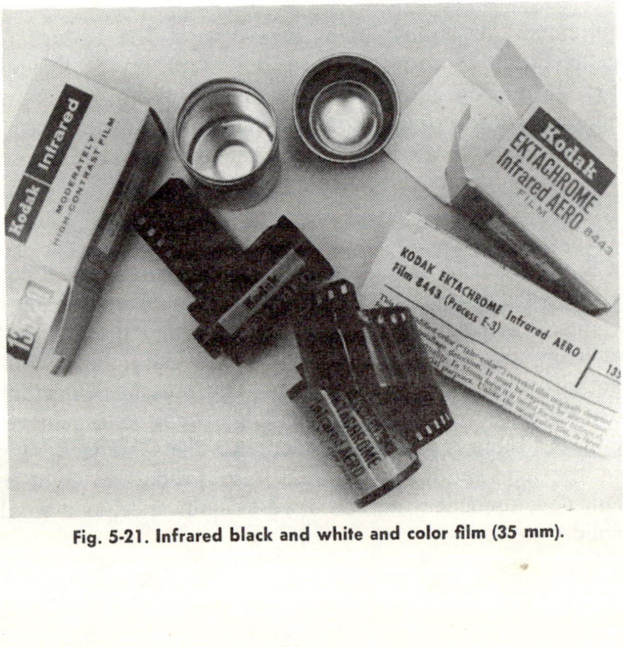

Fig. 5-21. Infrared black and white and color film (35 mm).

On black and white infrared film an IR source such as a GaAs LED appears in various shades of gray to white. If the diode has sufficient power and if the camera is focused properly, the diode will almost always look bright white. With the Ektachrome, infrared-emitting diodes and lasers appear as various shades of orange and red. Beam patterns can be photographed by illuminating a diffuse white card with the diode and photographing the resultant pattern, but the problem here is in focusing the camera properly. Since infrared has a longer wavelength than visible light, the lens must be moved slightly farther away from the film than in normal visible light. This technique is also hindered by the LED coming between camera and white card, and an alternative approach is to shine the LED against a ground-glass screen and photograph the patterns from the opposite side of the screen. Unfortunately, resolution of the infrared pattern will not be ideal with this technique.

The best way to photograph infrared patterns from LEDs is by direct illumination of the photographic film. The easiest technique is to load the film in a conventional camera and remove the lens. The LED is placed near the film plane and the shutter is activated with the camera in total darkness. This technique works best with cameras having a focal-plane shutter. Other cameras having blade shutters in the lens housing can be used but only in total darkness and with the LED turned off until the actual exposure period.

The exact length of exposure will vary from a fraction of a second to many seconds, and the only way to determine the best exposure time is by experiment. This method of extended exposure will yield a usable negative from an IR beam that would not necessarily be visible with expensive image-converter tubes. The infrared photos in Chapter 2 were made by biasing the LEDs at 10 mA, placing them about a centimeter from the film, and activating them for a small fraction of a second.

There are many other valuable hints and techniques for using infrared-sensitive film. For example, a yellow filter should be used over the camera lens if the infrared film is used in the presence of both infrared light from an LED and normal ambient light. This is because the infrared-sensitive layer is also sensitive to blue light. The yellow filter will effectively block the blue and permit the infrared to expose the film.

CHAPTER 6

Optics for LEDs and Detectors

The use of various optical techniques can almost always enhance the operation of both visible and infrared LEDs. Most LEDs incorporate a lens and perhaps a reflector within their mounting package and appear to have no need for external optics. However, these self-contained optics, particularly those of infrared emitters, are inferior to what can be achieved by means of a single external lens.

Lenses are not the only optical devices that can be used to good advantage with LEDs. Various kinds of reflectors, fiber optics, filters, and other devices can also be important. In this chapter many optical devices are considered in detail. But first there is a detailed discussion on the various optical characteristics of LEDs themselves. This discussion is important since it is not possible to properly specify external optics for LEDs unless the emission characteristics of the LED are known and understood.

POINT-SOURCE EMITTER

If a source of light emits light uniformly in all directions and has a maximum dimension less than 10 percent of the distance from the source to a detector, it is defined as a *point source*. Most LEDs can be treated as point sources in practical applications. It is possible to specify the amount of light from the source that strikes a known area if the distance, R, from the source to the illuminated area is known. Table 6-1 lists several relationships and definitions that apply to a point-source emitter in radiometric terms, and Fig. 6-1 shows the geometrical relationships for a point source and the equations in the table. A sample calculation will show the value of the table in determining the power density at an area a specified distance from the source.

Assume a hemispherical emitting GaAs:Si LED with a radiant output of 20 mW. Since the LED emits into a hemisphere, the diode has

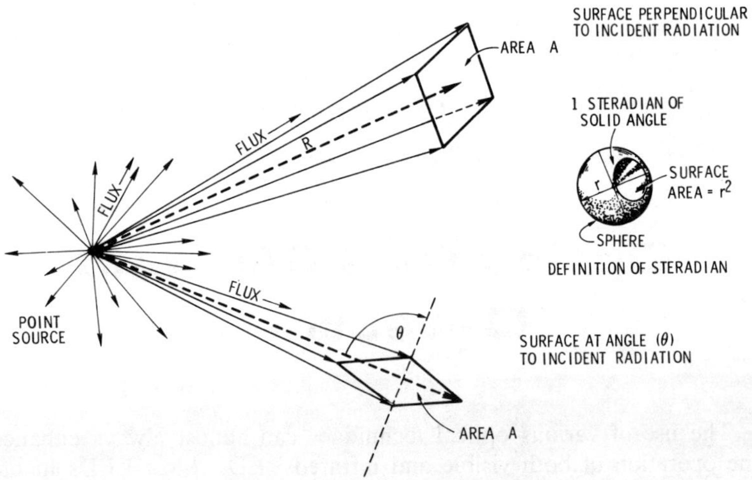

Fig. 6-1. Point-source relationships.

an intensity of 20 mW per 2π steradians (remember that a sphere has 4π sr). This value corresponds to a value of 0.020 w/2π sr or 0.003 w/sr. If the area being illuminated is a detector with an active area of 1 cm² exactly 10 cm from the LED, we can use the formula for incident flux density for a receiver of a known size in Table 6-1, Equation 6-4, to calculate the power received by the detector. Flux density equals the output of the LED in watts per steradian times the receiver area divided by the square of the distance to the receiver. Therefore, the detector receives a total power of 0.00003 watt per square centimeter or 30 microwatts.

Table 6-1. Radiometric Point-Source Relationships

Item	Equation	Equation No.
Source intensity (I)	watts/steradian	6-1
Incident flux density (E)	$\frac{I}{R^2}$ watts/sq unit	6-2
Incident flux density at angle (E)	$\frac{I}{R^2}\cos\theta$ watts/sq unit	6-3
Flux at receiver of known area (P)	$\frac{IA}{R^2}$ watts	6-4

This procedure holds, so long as the distance from the source to the detector is greater than ten times the source diameter. Since the diameter of the hemispherical source required to produce 20 mW is only 36 mils (0.09 cm), the point-source relationship holds until the detector comes within about a centimeter of the LED.

Table 6-1 can also be used to calculate values of incident power for a variety of situations. If the detector is tilted with respect to the source so that it presents less than its maximum cross section, the flux-density formula is simply modified by multiplying the result by the cosine of the displacement angle. The result is Equation 6-3.

EXTENDED-SOURCE EMITTER

A source of radiation with a maximum dimension greater than 10 percent of the distance to a receiver is said to be an *extended* or *area source.* An area source can have the property of emitting light whose intensity at any point is proportional to the cosine of the angle between the point of interest and the emission angle for maximum intensity. If the area source emits in such a manner, it is said to be a perfectly *diffuse* or *Lambertian* emitter. It is important to note that when a spot of light is directed onto a white card, the card itself becomes an area or extended source of illumination even though the card did not originate the light. This simple fact can be of great value in calculating the light received by a detector adjacent to an LED in a system designed to detect objects by their reflection.

Most natural and man-made objects have approximately diffuse reflection properties, but most have a degree of *specular* reflectance as well. A perfect specular reflector reflects incident light without distorting the beam divergence. A mirror is a good specular reflector. Most objects are not as specular as a mirror, but many have shine or gloss characteristics that give them both diffuse *and* specular reflectance.

Fig. 6-2 shows a representation of an area source and a detector, and the various radiometric relationships for such a source are given in Table 6-2. Note that as the separation distance between the source and detector decreases, the power density at the detector increases until it approaches the total radiation power of the source. In fact, for most practical purposes a separation distance of less than 10 percent of the radius of the source results in the maximum flux density being incident on the receiver. If the receiver is as large as the area of the source, or larger, it will receive nearly all the radiation of the source.

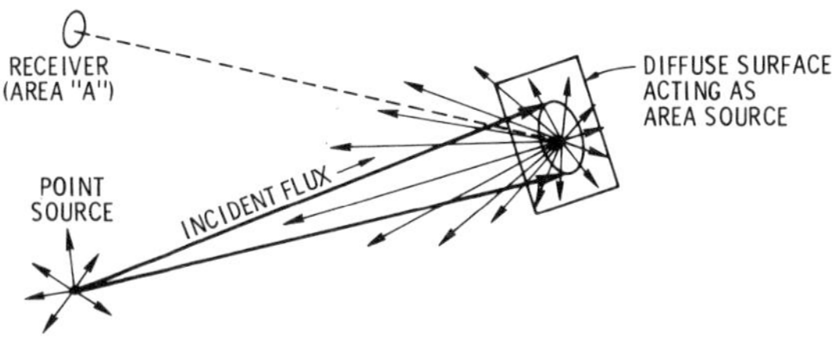

Fig. 6-2. Area-source relationships.

Table 6-2. Radiometric Area-Source Relationships

Item	Equation	Equation No.
Source intensity (L)	watts/cm²/steradian	6-5
Emitted flux density (M)	π L watts/cm²	6-6
Incident flux density (E)	$\frac{L\pi r^2}{r^2+d^2}=\frac{LA_s}{r^2+d^2}$	6-7
Total flux output (P)	MA_s	6-8

LED SOURCES

As was noted in the discussion on point sources, hemispherical LEDs can closely approximate a point source uniformly radiating into half a sphere. Though an actual point source radiates uniformly in all directions, the fact that a domed LED emits uniformly into half a sphere means that its radiant output can be approximated by point-source relationships.

LEDs with immersed optics (epoxy bubble lenses and reflectors) and external lenses may behave like either point sources or area sources under various viewing conditions, but it is usually safe to use point-source geometry for most LEDs. Diodes whose output radiation is collimated into a narrow beam will have characteristics of their own. In the latter case, a perfectly collimated LED beam will possess the same power density at any distance from the source. If the beam has divergence, power density can be calculated at any distance from the beam if the peak power of the diode, the beam divergence, and the distance are known. Fig. 6-3 shows several LED radiating possibilities.

Calculations of power density from an LED at a specified distance from the diode can sometimes be tricky. For example, consider a diode emitting 10 mW into a 10-degree beam. The power density at a point 10 cm from the diode will be 10 mW per beam area. The beam area is determined by using simple trigonometry to find the radius of the illuminated circle 10 cm from the diode. The radius of the circle equals the tangent of half the divergence times the distance, or 0.875 cm. The area equals π times the radius squared or 2.4 cm². Therefore the power density 10 cm from the LED is 10 mW per 2.4 square centimeters or 4.17 mW per square centimeter. This procedure can be simplified with the following expression.

$$E = \frac{4P}{\pi\theta^2 d^2} \qquad (6\text{-}9)$$

where,

E is the power density in watts per square centimeter,
P is the LED total output in watts,
θ is the beam divergence in radians,
d is the distance in centimeters.

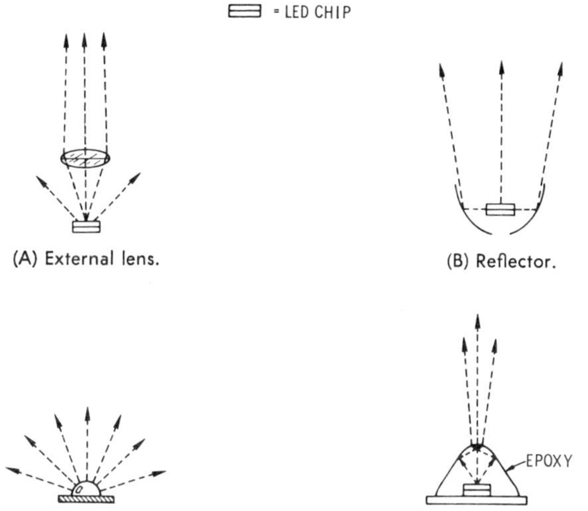

Fig. 6-3. LED radiating configurations.

Equation 6-9 simplifies matters since there is no need to look up the tangent of the angle of divergence. The angle is simply converted from degrees to radians by multiplying the number of degrees by 0.01745. Thus 10 degrees equals 0.1745 radians, and 57.3 degrees equals 1 radian. Substituting the values given above into the equation, we confirm a power density of 4.17 mW/cm^2.

APPLYING POINT- AND AREA-SOURCE RELATIONSHIPS

This discussion of point- and area-source relationships can best be summarized by two general rules:

1. Use point-source geometry when calculating the optical power from an LED at a distant surface.
2. Use area-source geometry when calculating the optical power reflected from a diffuse target.

Point-source geometry is particularly useful when calculating the parameters for an optical communications link, intrusion alarm, or broken-beam detection scheme. Area-source geometry is valuable for calculating the detection range of an optical radar against diffusely reflecting targets. Of course, both geometries have numerous other uses. For example, area-source geometry can be used to calculate the amount of sunlight reflected from background objects in the field of view of the receiver in an optical communications link. Since exces-

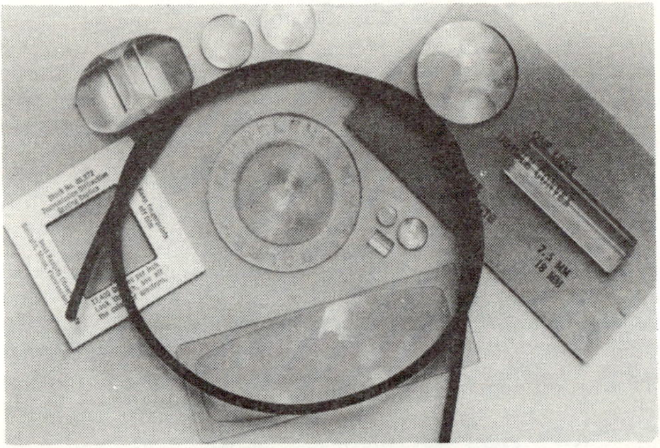

Fig. 6-4. Assortment of optical devices.

sive sunlight can reduce receiver sensitivity, the ability to predict the amount of sunlight for a variety of conditions can be valuable.

All of the relationships in Table 6-1 and 6-2 are given in radiometric units since infrared LEDs are most frequently employed in applications requiring point- and area-source calculations. For photometric applications, simply insert the appropriate photometric terms into the equations.

LENSES

A lens is a transparent device that employs its geometry and refraction to manipulate an incoming beam of light. Lenses may be made of glass, plastic, or any material that will hold its shape and appear transparent to the wavelength of interest. A lens can be as simple as a droplet of water or as complex as a precision ground slab of infrared-transmitting germanium.

Glass lenses are generally expensive. Recently there has been much interest in commercial plastic lenses. Costing only a fraction of their glass counterparts, plastic lenses are lightweight and shatterproof. Though their optical quality is not as impressive as glass lenses, plastic units are generally adequate for use in LED source and detector systems. One advantage of plastic in detector-lens applications is that an infrared-transmitting red dye can be easily added to the lenses during manufacture to act as an integral filter for ambient light. The use of epoxy lenses as immersed optics for individual LEDs and arrays has been mentioned several times earlier in the text. A photographic enlargement of an array of plastic bubble lenses is shown in Fig. 6-5.

Several categories of lenses can be of great value when used with LEDs and detectors. The most useful is the *converging* or *positive* lens.

Fig. 6-5. Enlargement of plastic-bubble lens array.

This kind of lens has one or both surfaces curved outward in a *convex* or *double convex* configuration. A positive lens accepts a beam of parallel light and causes it to be focused at a spot a certain fixed distance away called the *focal point*. Conversely, light emerging from a spot at the focal point will be converted into a parallel beam by the lens. *Diverging* or *negative* lenses spread out an incoming light beam and therefore have a negative focal length.

The relationship of a lens to its focal point is shown in Fig. 6-6. Many LED applications employ positive lenses, and the ability of a positive lens to collimate divergent radiation from a point source into a narrow-divergence beam is shown in Fig. 6-7.

Simple lenses can be made from a single piece of glass or plastic, or two or more separate lenses can be bonded together to give a *compound lens*. The compound lens has the characteristic of focusing light rays of differing wavelengths to a common point. Because different wavelengths receive differing degrees of refraction, a simple lens will not focus all wavelengths to the same point. The effect is illustrated in Fig. 6-8.

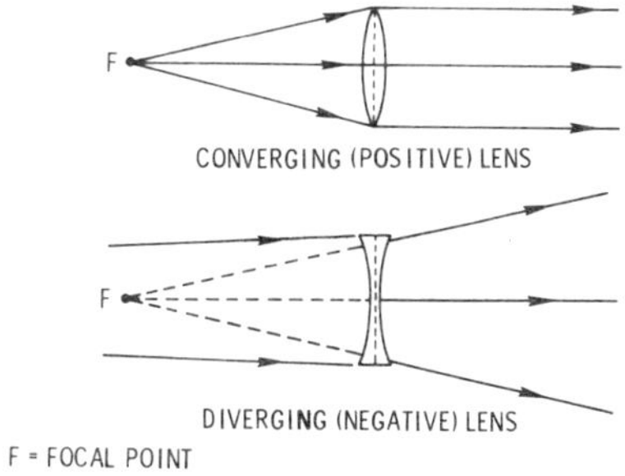

Fig. 6-6 Focal points of converging and diverging lenses.

Compound lenses are frequently called *achromats* or *color-corrected lenses* because of their desirable focusing characteristics. Two bonded simple lenses make up a *doublet,* while a triplet incorporates three separate lenses. Transparent cement such as Canada balsam is used to bond the lenses into the compound arrangement.

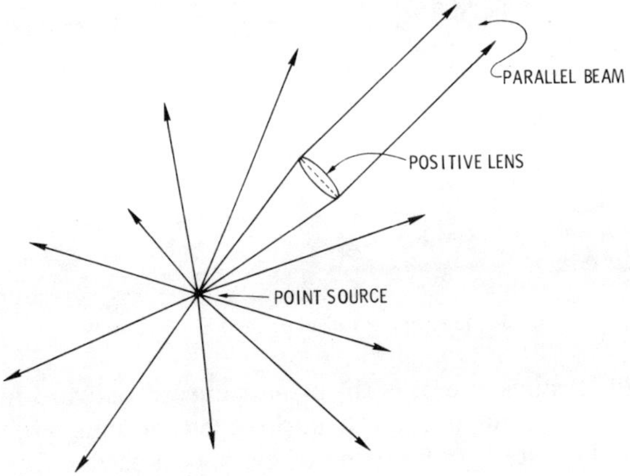

Fig. 6-7. Collimation of light with a simple lens.

Since LEDs have relatively narrow spectral widths, it is not necessary to employ achromats for most applications. For optimum results in precision applications, however, achromats are a must. They are more expensive than simple lenses, but they can significantly reduce the beam divergence of an LED-lens assembly. Conversely, they can focus light from an LED into a smaller point than a single simple lens.

(A) Simple lens. (B) Color-corrected compound lens.

Fig. 6-8. Color distortion and correction.

An example of the effectiveness of an achromat is shown by the beneficial results had with their use in diode-laser transmitter systems. Military illuminator and communication systems using diode lasers frequently employ color-corrected achromats for optimum beam collimation and focusing, even though the spectral width of the laser may only be 4 nanometers or so.

Generally lenses have a spherical surface, but variations exist that can be very useful. Fig. 6-9 shows a cylinder lens. This device bends light rays into a linear focal region parallel to the plane of the lens.

Cylinder lenses can be very useful with linear detectors and detector and emitter arrays with one dimension longer than another. They are also useful in enlarging LED displays.

Whether used with an LED or a detector, a lens can be thought of as a collection device. If it gathers radiation into an organized beam with higher power density than a widely divergent beam, it has *optical gain*

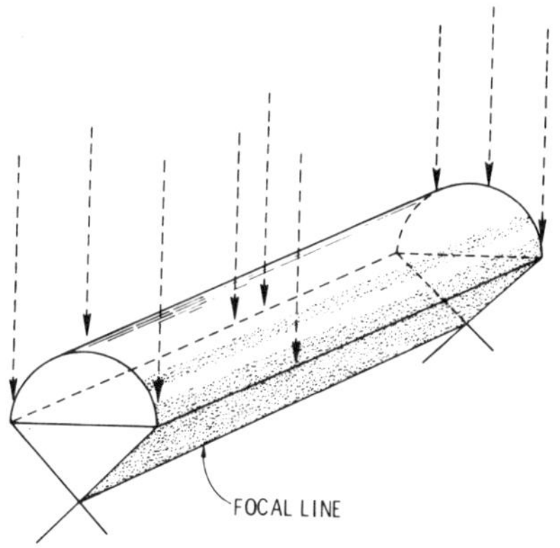

Fig. 6-9. Cylinder lens.

just like some kinds of radio receiving antennas. The gain is not obtained without some cost, since some radiation will be reflected outward at both the front and back surface of a lens, and absorption will substract a little more. As a result, a typical lens may have a transmission of from 90 to 96 percent. Coating with thin antireflection (ar) films will decrease reflection losses, but even with 10-percent loss, most lenses add a substantial optical gain to optoelectronic systems.

USING LENSES

There are two major uses for lenses in LED photodetector systems:

1. Collimating the divergent radiation from an LED chip into a narrow beam.
2. Increasing irradiance on a detector by collecting more of the available radiation in the vicinity of the detector. By a judicious use of simple lenses, it is possible to maintain the maximum desirable irradiance on a detector over extremely long distances.

The most important parameters for a lens are diameter and focal length, and by means of a simple formula it is possible to calculate the beam divergence that will result from pairing a light source with a lens:

$$\theta = \frac{d}{f} \tag{6-10}$$

where,

θ is the divergence in radians,
d is the width of the source,
f is the focal length.

Both d and f must be expressed in the same units of measurement. As a sample calculation, assume an LED with a chip 0.1 mm in diameter and a lens with a focal length of 30 mm. The divergence would then equal only 0.003 radian. The answer can be converted to degrees by multiplying by 57.3 (0.003 radian equals 0.17 degree).

Most LEDs have a self-contained glass lens or immersed epoxy optics that can complicate matters. The LED lens in effect becomes the source, and, because of its larger d, beam divergence increases considerably for the same conditions. Referring again to the previous example, assume that the 0.1 mm LED chip is immersed in a blob of epoxy 4.0 mm across. Substituting the values into Equation 6-10, we find that the lens of 30-mm focus gives a divergence of 0.133 radian (7.6 degrees), about 44 times greater than the LED chip without the epoxy lens. This situation may not be as bad as it seems, for, as usual, there is an important tradeoff to consider. The chip alone may produce a much broader radiation pattern than a properly shaped epoxy lens. Therefore, the arrangement where radiation from the chip alone was collimated may well have been very inefficient since much of the radiation may never have reached the lens. While the epoxy lens causes a larger divergence, if properly shaped it directs much more radiation from the chip to the collimating lens. The end result is a beam with more total power than the initial beam.

The trading-off process continues when the power density per unit area contained within each beam is measured. Because the second beam has such a broad divergence, its power density may be only slightly higher or even less than that of the beam from the unencapsulated chip. Since the beam pattern and lens size of LEDs vary so widely, the best procedure for collimating the radiation from a particular diode is comparison of the data sheet information about the beam pattern with desired divergence and allowable lens dimensions.

It should be evident by now that the smaller the source being collimated by a particular lens, the narrower the divergence. It is also evident that the lens diameter is directly related to the amount of radiation it collects. Ideally the diameter of a lens should be as large as possible in order to collect all the emitted light from an LED. The figure of merit for collection is the f/number, sometimes referred to as the "speed" of a lens. The f/number of a lens is found by a second basic lens formula:

$$f/\text{number} = \frac{f}{d} \qquad (6\text{-}11)$$

where,

f is the focal length,
d is the lens diameter.

An f/1 lens has a focal length equal to the diameter and is considered a "fast" lens, and a lens with an f/number as high as 16 is considered a "slow" lens. The f/number of a lens can be easily altered by means of a mechanical aperture that reduces its effective diameter. This is how a good camera lens is changed from a fast f/2.5 or so to a slow f/22.

It is technically difficult to make lenses with f/numbers less than unity, but by considering the parameters of a typical LED it can be seen that a lens with a small f/number is desirable for collecting as much emitted radiation as possible. As a general rule, an f/1 or somewhat "slower" lens can be employed, but if it is important to preserve a small beam divergence, a lens with a focal length at least ten times the diameter of the LED should be selected. For an f/1 lens this obviously means a bulky optical system.

Fortunately there is an easy way to preserve low beam divergence while collecting all or most of the radiation from an LED. Several companies make infrared GaAs:Si LEDs with a miniature integral parabolic reflector and immersion epoxy optics. The reflector serves to catch the edge emission from the diode chip and focus it outwards. The epoxy reduces the critical interface problem by providing a material with a low index of refraction between the semiconductor crystal and air.

The reflector alone causes these diodes to project beams with a divergence of about 30 degrees. Since the diameter of each diode reflector is quite small (2.3 mm for the RCA SG1004 and 0.75 mm for the GE SSL-55C) a very compact and economical f/1 lens can be used to obtain a narrow beam containing most of the emitted radiation. In a system built by the author, the radiation from a GE SSL-55CF LED was collimated with an f/1.2 lens with a focal length of 14 mm. From Equation 6-10, the divergence of the resultant beam should have been 0.054 radian, or about 3 degrees. The actual divergence was 6 degrees, with the error being caused by the epoxy blob over the LED chip and reflector. The epoxy formed a circle with a diameter about twice that of the 0.75-mm diameter of the reflector. Using the epoxy blob as the source size and recalculating, the predicted divergence is found to nicely match the actual value.

A helpful technique in LED beam collimation concerns reducing source size to more nearly resemble a point source. The RCA diodes with the integral reflector are already quite small. GE's diodes containing an integral reflector are available with a flat glass lens or with a collimating lens. If it is desired to achieve a smaller divergence than the LED plus integral glass lens offers (about 20 degrees), carefully cut off the top of the metal can containing the lens with a rotary grinder or jeweler's file. This permits an external lens to be employed with little difficulty. The epoxy coating over the LED chip will protect it from contamination, but use care to avoid damaging the chip. Ideally, use a flat-windowed version of the diode.

Many kinds of epoxy-covered diodes can be artifically reduced in size by *careful* grinding of the plastic coating. Great care must be used to avoid damaging the LED chip and its connection wires. This same technique can be used to vary the beam shape of epoxy-coated emitters. One technique is to use fine sandpaper and then polish the ground plastic with a fine abrasive like scouring powder and toothpaste. Commercial epoxy polishing materials are available.

LENS ALIGNMENT

Proper lens alignment is important when using a lens with a very small point-source emitter. This is particularly true with diode lasers, since a misalignment of a fraction of a millimeter can cause the output beam to be badly formed and to have excessive divergence. Two infrared photos showing properly and improperly aligned diode-laser patterns are given in Fig. 6-10. A properly aligned LED and lens is represented by the beam pattern in Fig. 6-11. Both laser and LED photos were made with infrared film. Note the excellent uniformity of

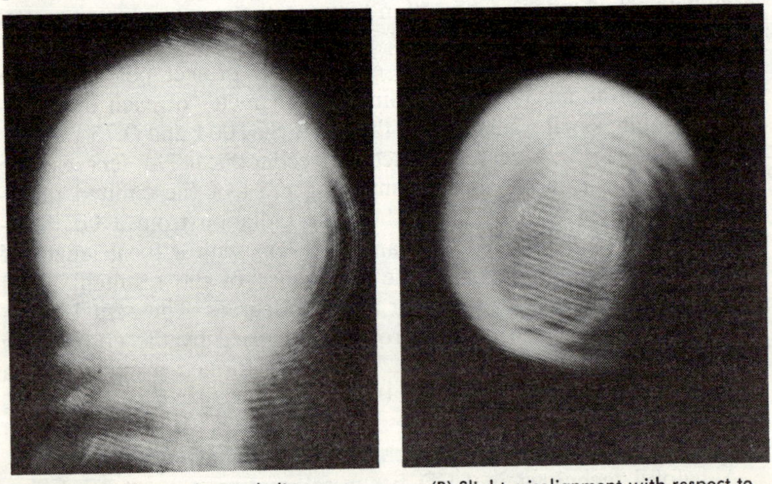

(A) Relatively good optical alignment with lens.

(B) Slight misalignment with respect to focal point of collimating lens.

Fig. 6-10. Diode laser alignment.

the LED beam pattern when compared to the laser photographs. The latter include fine striations due to interference caused by the lens and the quasi-coherent nature of the laser radiation.

The diode laser is more difficult to align than the typical LED since the latter has a much larger source size. Nevertheless, LED lens alignment should be accomplished with appropriate care.

The use of an image-conversion device is of great help when aligning infrared LEDs in a lens system. Infrared phosphor screens or image-converter tubes provide a clear indication of beam shape, size, and uniformity. Alternatively, a possible solution is to employ a green-phosphor-coated upconverter LED and collimate the green emission with the external lens. When the green is aligned, the infrared will be nearly aligned also. The problem here is that infrared has a longer wavelength than visible green. Therefore, the lens may need to be moved slightly farther away from the diode to achieve a proper infrared focus.

Another kind of lens misalignment can occur when a lens is used with diodes having an integral glass or epoxy lens. As shown in Fig. 6-12A, if the lens is not focused properly, it may offer little or no optical gain over the diode's own lens. This is particularly true with detectors. A good way to align an external lens and a diode containing an integral lens is to look into the external lens from a distance of half a meter or more while focusing. When the lens appears to be completely filled with the black surface of the detector, it is at or near proper alignment.

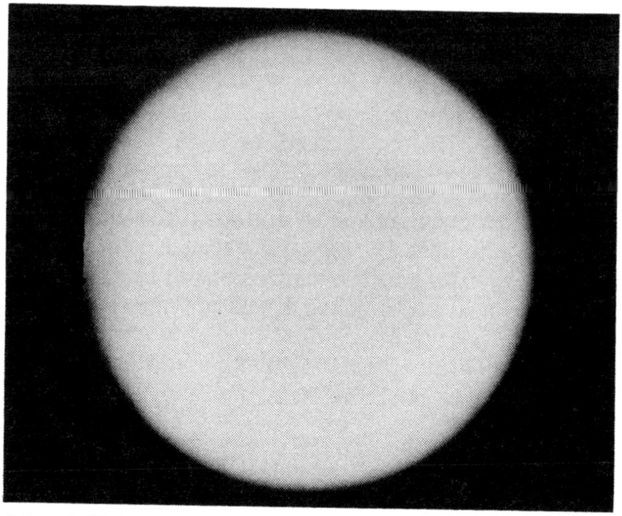

Fig. 6-11. Infrared LED in good optical alignment with external lens (note beam uniformity as compared to laser beam patterns).

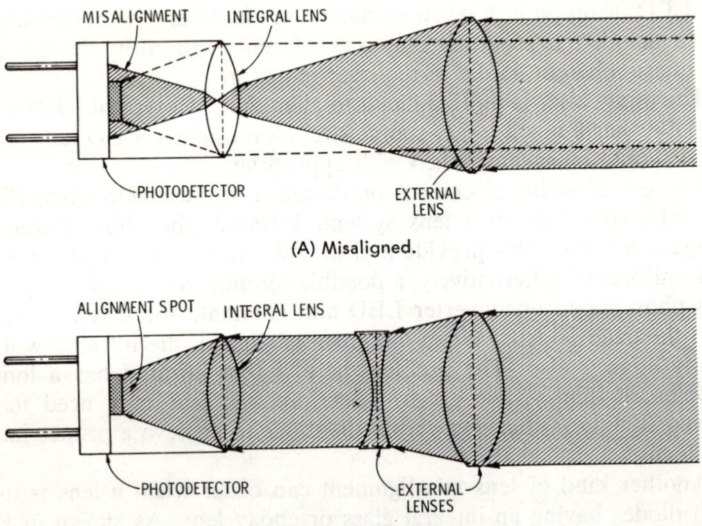

(A) Misaligned.

(B) Properly aligned.

Fig. 6-12. External lens alignment.

Fig. 6-12B shows a method for improving collection efficiency by employing a second, negative lens in addition to the original positive lens. In this arrangement the negative lens converts the incoming beam from the positive lens into parallel rays, and this permits the integral lens of the diode to achieve proper focus. The disadvantage of this arrangement is added complexity, cost, and losses. A preferrable alternative is to remove the integral lens and substitute an external lens of the desired dimensions.

UNUSUAL LENS EFFECTS

An LED chip has a finite shape and area, and therefore a lens will project a magnified image of the source when the chip is at the lens focal point. The result can be a square beam for some diodes. Diodes with an integral miniature parabolic reflector will be projected as either a uniform round beam or a round beam containing a square insert. If a square shape is undesirable, the lens can be displaced slightly to blur the square shape into a round outline, but this procedure may also increase the divergence of the beam.

PARABOLIC REFLECTORS

As has been previously noted, miniature parabolic reflectors have been employed with good results as a part of the integral optical system of some LEDs. Because of the emission characteristics of many LEDs, larger external reflectors can also be used to advantage.

Perhaps the most dramatic example of where an external parabolic reflector enhances performance of an LED is the domed emitter. Texas Instruments manufactures a variety of external reflectors for several of its high-powered, domed emitters. The reflectors are gold plated to enhance their reflectance and they emit beams with a divergence of from 10 to 40 degrees, considerably less than the normal 2π steradian emission pattern from these diodes. While these commercial reflectors are designed for mounting the LED, a conventional metal flashlight reflector can also be employed. The reflector will have to be appropriately modified to accept the LED.

An economical way to obtain an efficient, high-quality reflector is to remove the glass window from a sealed-beam headlight. Care should be used to avoid personal injury and damage to the reflective coating on the inside of the lamp. The LED can be mounted to the existing filament supports inside the lamp, but before it is fixed in position the diode should be placed at the focal point of the reflector. One way of determining when the diode is at the focal point is by observing the reflector from a position half a meter or more away. When the entire reflective surface of the reflector is filled with the image of the black surface of the emitter, the diode is properly aligned.

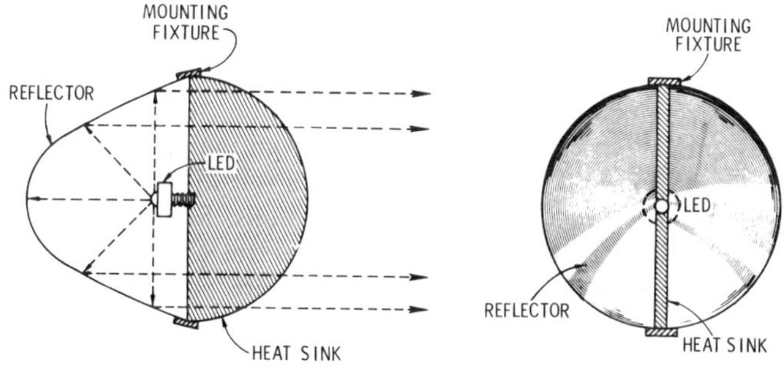

Fig. 6-13. Parabolic reflector plus heat sink.

Operation of a high-power LED requires heatsinking for maximum power output, and Texas Instrument's custom parabolic reflectors serve a double function as heat sinks since they are in intimate contact with part of the diode package. One way to achieve heatsinking for a diode mounted in a conventional reflector or sealed-beam lamp is shown in Fig. 6-13. The semicircular plate protruding from the reflector provides a good heat sink while blocking only a tiny fraction of the emitted beam. This kind of heat sink can also be designed to provide a sturdy, adjustable mount for the LED.

FIBER OPTICS

Fiber optics can prove extremely useful in numerous LED applications. With their ability to transmit a light beam from one point to another, optical fibers can be used in optical communications, remote switching, signal coupling, and numerous other applications. One unique application in use for several years now employs an LED and a fiber-optic cable to transmit signals from a hospital patient to a remote monitoring unit. The system prevents the possibilities of dangerous electrical shock inherent in conventional patient monitors with electrical connections.

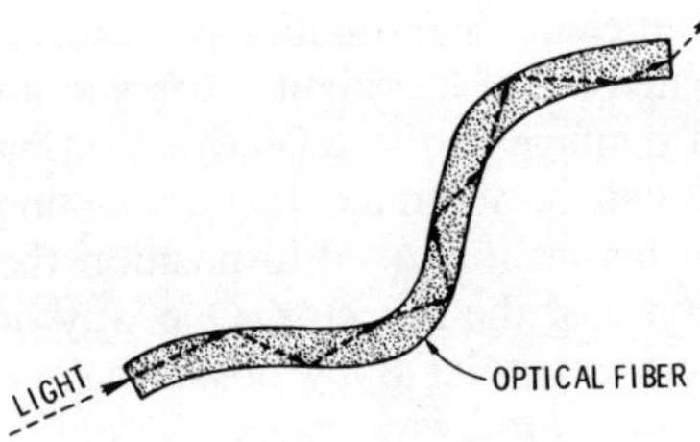

Fig. 6-14. Light transmission through optical fiber.

Sometimes called a "light pipe," a fiber-optic strand is actually an *optical waveguide*. Transmission through the fiber is demonstrated in Fig. 6-14. After a light ray enters the fiber it eventually strikes the fiber/air interface. The difference in index of refraction between the fiber and the outside air causes the glancing ray to be reflected back into the fiber rather than allowing it to exit into the air. The process continues until the ray ultimately emerges from the opposite end of the fiber, with the major losses being due to absorption within the fiber itself and some leakage through the fiber walls. Sometimes reflection within a fiber is enhanced by cladding a central core with a material having a lower refractive index. Clad fibers are particularly desirable when optical "crosstalk" between closely spaced fibers must be prevented.

Fig. 6-15 shows the end of a fiber-optic cable transmitting sunlight from a point several feet away. Some of the fibers are darker than others because of uneven illumination on the sunlit end.

In a fiber with parallel sides, light striking one end will be conducted into the fiber within an angle of acceptance governed by the following equation:

$$\sin \theta = \sqrt{n_1^2 - n_2^2} \qquad (6\text{-}12)$$

where,

θ is half the acceptance angle,

n_1 and n_2 are the indexes of refraction for the fiber and air (or the cladding), respectively.

Equation 6-12 gives a figure of merit for the fiber acceptance angle, the *numerical aperture* (na). An expression of light collecting power, na corresponds to the sine of half the acceptance angle.

Optical fibers may have an na ranging from 0.2 to almost 1.0. The na of a lens is expressed by

$$\text{na (lens)} = \frac{1}{2\text{f/number}} \qquad (6\text{-}13)$$

An f/1 lens has an na of 0.5, while slower lenses have progressively smaller values.

The transmittance of light from an LED through an optical fiber is affected by the na, losses within the fiber, and mechanical placement of the fiber with respect to the diode. If the fiber is placed too far away from the LED chip, much of the light will not come within the receiving aperture of the fiber, and if the chip is larger than a single fiber, a cable of several fibers may be necessary.

The radiation density transmitted by a fiber can be expressed:

$$E = L(\text{na})^2 T^r \text{ watt/cm}^2 \qquad (6\text{-}14)$$

where,

E is the radiation density,
L is the radiance at the input of the fiber (corresponding to the radiance of the source if the two are in intimate contact),
T is transmission of the fiber per unit length,
r is the fiber length in the same units of length.

The advantage of using a fiber-optic transmission technique can be seen by comparing an LED-detector arrangement with and without a fiber. For simplicity, assume that all of the emitted radiation falls within the acceptance angle of the fiber. Therefore, the na is 1.0. If the LED source has a radiant intensity of 15 mW/steradian and the separation distance from LED to detector is 20 cm, then the incident flux

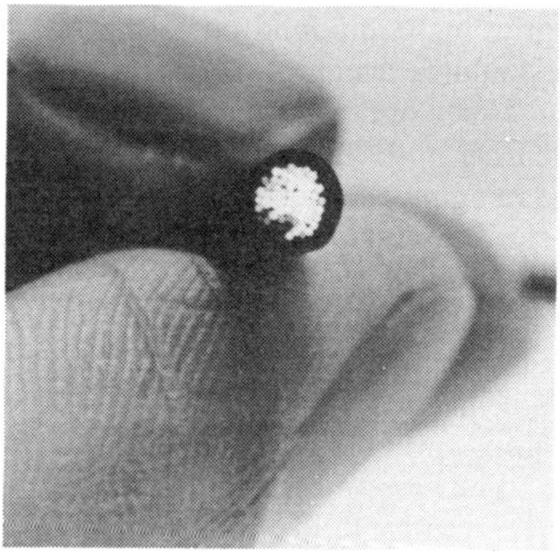

Fig. 6-15. Light emerging from fiber optical bundle.

density at the detector can be calculated from Equation 6-2 in Table 6-1:

$$E = \frac{I}{R^2} \tag{6-2}$$

$$= 3.75 \times 10^{-5} \text{ W/cm}^2$$

Assuming a fiber that collects 15 mW of the radiation and has a transmission of 0.9 per 10 cm, Equation 6-14 gives a transmitted irradiance of

$$E = L\ (na)^2 T^r \text{ W/cm}^2 \tag{6-14}$$

$$= 1.22 \times 10^{-2} \text{ W/cm}^2$$

The fiber-optic transmission technique in this sample problem is obviously superior, since the fiber supplies an irradiance 325 times that of the emitter without the fiber. Lenses and other optical systems can provide similar coupling efficiency, but the fiber-optic approach is particularly convenient, simple, and low in cost.

Fiber-optic strands and cables are available from original manufacturers as well as dealers such as Edmund Scientific Company, Radio Shack, and others. The infrared transmission of plastic fibers is not as good as that of glass fibers, but glass is generally more costly. Fibers can be purchased in bulk coils and assembled into cables by inserting individual strands into heat-shrinkable tubing and heating the tubing to form a tight fit around the fibers. The melting point of plastic fibers must not be exceeded during the shrinking operation (175° F for the widely available DuPont Crofon fibers).

A method of making an even thicker cable of fibers is to wrap a continuous loop around two nails placed a desired distance apart. The fibers are secured with tape and the looped ends are cut. The ends of plastic fibers can be polished by first cutting with a razor blade to give a smooth, flat surface and then polishing with wet No. 600 alundum grinding compound on a smooth glass or metal plate. Final polishing should be accomplished with jeweler's rouge.

Fiber-optic strands can be coupled directly to plastic encapsulated LEDs by boring a hole into the top of the diode. Use care to avoid damaging the delicate LED chip assembly. A single fiber or several can be inserted into the hole and secured in place with clear epoxy. Epoxy can also be used to secure fibers adjacent to nonencapsulated diodes. In the case of diodes installed in a transistor header, the lens or window is removed and the can is filled with clear epoxy. The fibers are inserted into the epoxy, placed as close as possible to the LED chip, and held in place until the epoxy hardens.

For more information on fiber optics, consult references on the subject and write distributors and manufacturers. Since costs are coming down and transmission characteristics are being improved, fiber optics should be given careful consideration in LED-detector coupling arrangements.

FILTERS

The previous chapter noted that many detectors are saturated by high ambient light levels. An excellent way to avoid this problem is to use an optical fiber that passes only the desired wavelength while blocking most others. The filter will absorb some of the desired wavelength also, but the overall benefits will far surpass any losses.

Filters used in most LED-detector applications are usually of the absorption or interference variety. The former consists of glass or plastic containing or coated with a substance relatively transparent to the wavelength of interest. Absorption filters are inefficient since they may pass only half of the desired wavelength and because they pass a band of wavelengths generally wider than the emission spectra for most LEDs. Their chief advantage is low cost.

Fig. 6-16. Optical Coating Laboratory infrared interference filter.

Interference filters make use of multilayer dielectric layers to reflect unwanted wavelengths while passing a narrow band of desired wavelengths. The layers consist of very thin metal films separated by precise thicknesses of inert material. Since the layers must be applied with care and precision, interference filters are far more expensive than absorption types. Frequently the benefits are worth the added cost, however, since interference filters can pass more than 90 percent of the desired wavelength and have a spectral bandwidth of only a few hundred nanometers or less. Fig. 6-16 shows an interference filter made by Optical Coating Laboratories, Inc. The filter has a transmission peak very close to the peak emission wavelength of GaAs LEDs. The spectral transmission characteristics of the interference filter and a typical absorption filter are compared in Fig. 6-17.

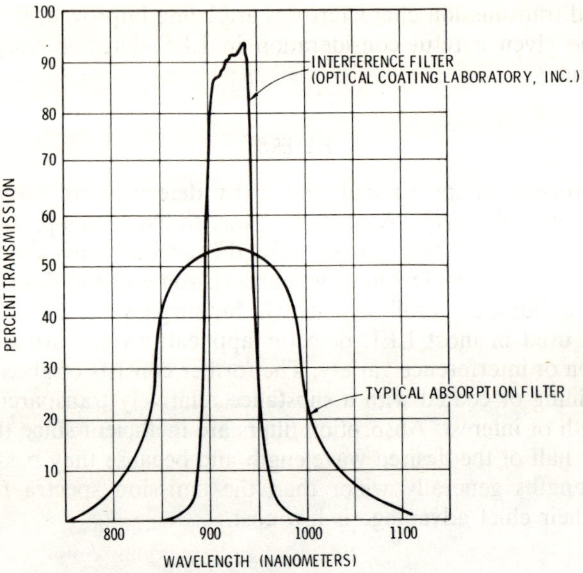

Fig. 6-17. Interference and absorption filter spectral transmission.

Some interference filters employ an absorption filter as a substrate. While this technique enhances overall rejection of undesired wavelengths, it results in high absorption of the transmitted light as well. An additional disadvantage is that these double filters are generally thicker than an interference filter alone.

CORNER REFLECTORS

A very useful device in optical ranging and detecting systems is the corner reflector. Consisting of three plane reflectors mounted so their apex forms a right-angle corner (Fig. 6-18), the corner reflector has the ability to project an oncoming beam of light directly back to the source. The Apollo astronauts placed several arrays of corner reflectors at various points on the moon. By reflecting high-power laser

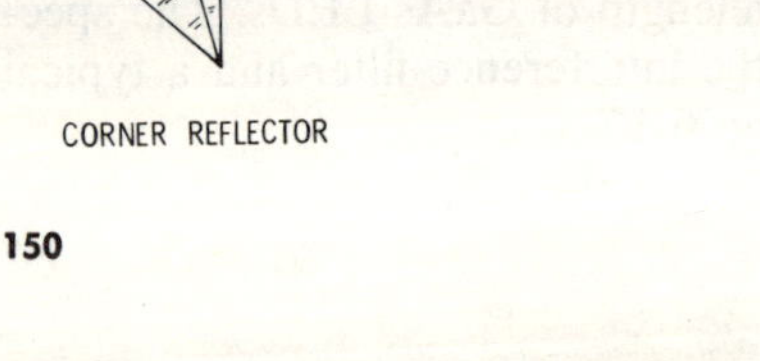

Fig. 6-18. Principle of the corner reflector.

beams from the reflector arrays and measuring the time of flight, scientists have been able to redefine with great accuracy the distance from earth to moon.

Corner reflectors are available in several configurations. Some are front-surface mirrors mounted in a corner configuration and assembled into a sturdy metal housing, and others are solid glass prisms. Both are generally costly, as the corner angles must be accurately aligned for the device to operate properly.

An inexpensive alternative to the precision corner reflector is the common plastic automobile or bicycle reflector. As the photographic enlargement in Fig. 6-19 shows, these devices are actually an array of hundreds of individual plastic cube corners. Use clear ("silver") plastic reflectors for best results with most visible and infrared LEDs.

Fig. 6-19. Enlargement of plastic corner-reflector array.

The *optical gain* of a plastic reflector array will be considerably less than that of a precision glass unit. Gain is a figure of merit that defines the reflection of a reflector over that of a perfect diffuse reflector. A plastic array may cause an incoming beam of parallel light to be reflected with a divergence of as much as a degree because of slight imperfections in the plastic, while glass corner reflectors may add a divergence of six arc seconds or less. As a result, a typical plastic reflector can have a gain of about 6000 over a diffuse reflector, while a good glass reflector can have a gain of 5,000,000,000.

Though plastic reflectors do not have the very high gain of glass units, their economy and ready availability make them well suited for short-range detection applications such as projecting an infrared beam from an LED across an entrance way to detect visitors. For convenience, all the electronics is housed in a single package and a reflector is used to return the transmitted beam to a detector. If the diffuse reflectance of the hallway alone were utilized, the power returned to the detector would be very low. A flat mirror would be a vast im-

Fig. 6-20. Enlargement of Scotchlite reflective tape.

provement, but it would need very careful alignment to keep the reflected beam on the receiver. A slight movement would probably disable the system. A small plastic reflector provides an ideal solution. Since the reflector does not have to be perfectly aligned with respect to the receiver, it will provide a good and inexpensive return device.

If available power is sufficiently high, reflective tape can be used in many close-range LED-detector applications. As shown in the photographic enlargement in Fig. 6-20, reflective tape is coated with a layer of tiny glass spheres that have the ability to retroreflect a light beam back to its source. Reflective tape is available in a variety of colors and configurations from automotive supply dealers.

BEAM SPLITTERS

A very useful device in some LED applications is the *beam splitter.* As shown in Fig. 6-21, the beam splitter allows part of an incoming beam to be transmitted while reflecting all the remainder. Splitters can be made to reflect from as little as 5 percent to as much as 95 percent of an incoming beam.

Beam splitters can be used to cause a single LED to project two or more separate beams. A second beam can be used actively or simply sent over to a reference detector designed to continually monitor LED output. Splitters can also permit an LED and a detector to share a

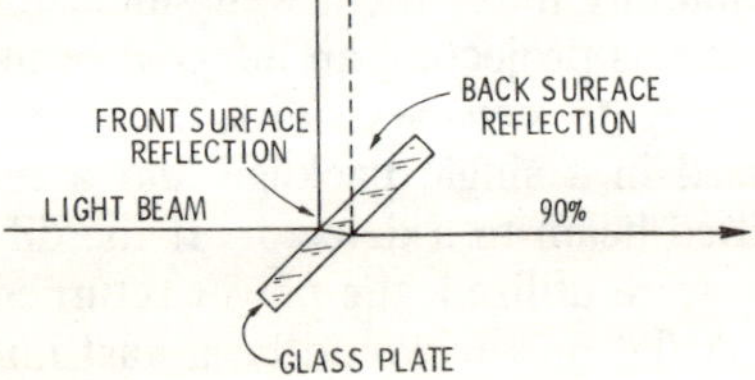

Fig. 6-21 Principle of a beam splitter.

common optical system, but some of the light (typically half) will be lost.

For very simple applications requiring a 10- to 90-percent splitter, a glass microscope cover glass or slide can be employed. For other ratios of reflected to transmitted light, Edmund Scientific and B&F Enterprises sell coated splitters for reasonable prices. In addition to half silvered "one-way mirrors," both companies sell quality prism beam splitters.

MIRRORS

Mirrors also find frequent use in optoelectronic applications. Inexpensive cosmetic mirrors can be used for many applications, but when it is desired to avoid secondary reflections from the glass front of a mirror, use *first surface mirrors* available from Edmund Scientific and others.

For unusual applications consider metal mirrors and various kinds of aluminized mylar. The former devices are inherently sturdy and can be used in environmental extremes (such as beam-turning mirrors in an outdoor LED intrusion alarm). Aluminized mylar can be used to make custom large-area reflectors and reflector arrays if it is cut into appropriate shapes and cemented to a metal, wood, or even corrugated board substrate.

BAFFLES, MASKS, AND SHIELDS

In many indoor, and some outdoor, applications a detector can be shielded from most ambient light by means of a simple baffle. Frequently all that is required is a hollow tube painted flat black on the inside. The tube can even serve to mount a detector lens.

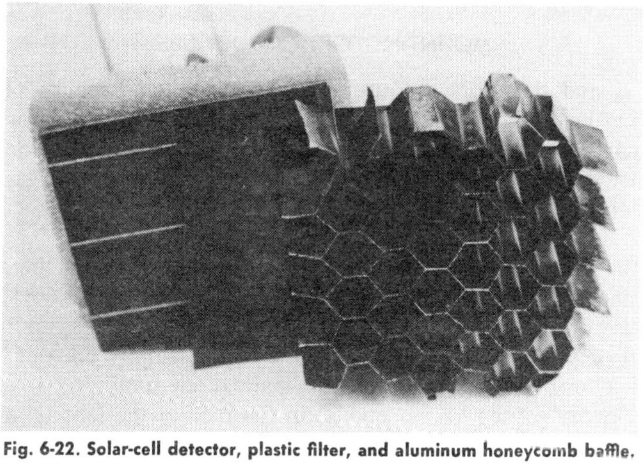

Fig. 6-22. Solar-cell detector, plastic filter, and aluminum honeycomb baffle.

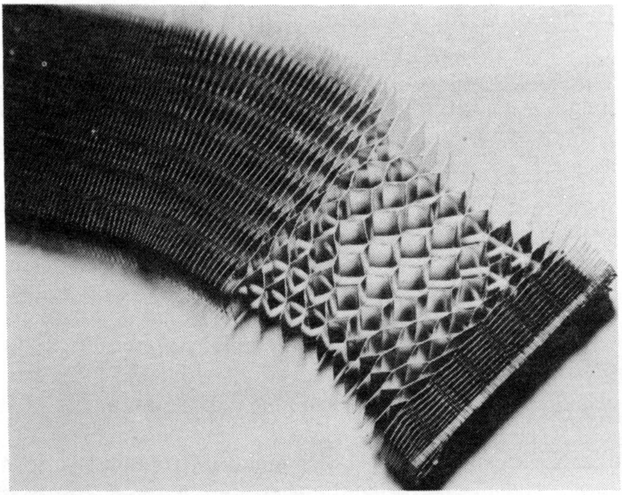

Fig. 6-23. Expandable aluminum honeycomb for use in optical baffles.

Fig. 6-22 shows a simple solar-cell baffle made from a soot-blackened piece of aluminum honeycomb. A length of honeycomb is shown in Fig. 6-23. The arrangement provides the receiving end of an experimental LED travel aid for the blind assembled by the author in 1966. Though the aid received rough handling by blind subjects in two years of tests, the baffle has not required a new application of soot.

Other baffle and shielding techniques are obvious. In close-range systems where a detector and LED are separated by only a few centimeters, the entire optical path can frequently be enclosed by an opaque metal, plastic, or even paper tube.

MOUNTING OPTICAL DEVICES

LEDs and detectors are usually easy to install because of their mounting leads, but optical components such as lenses, mirrors, and filters can prove very difficult to mount. Lenses are a particular problem since the very act of mounting can block part of the light-gathering area. Mirrors can at least be mounted by attaching a bracket to the unused side with epoxy.

Of the numerous techniques employed to mount lenses the author has found two to be ideal for most LED-detector applications. Both make use of a coaxial system wherein lens and diode are mounted in a metal or plastic tube. In the first, the lens is simply epoxied to the end of a tube that slides inside the main diode-mounting tube. Use care to avoid getting excess epoxy on the face of the lens; if a metal mount is used, cure the assembly in an oven at a few hundred degrees to promote fast drying. This technique permits the lens to be focused in and out.

A second technique makes use of rings of tubing that just fit inside the primary diode-mounting tube. Two rings are needed, one for each side of the lens.

In both mounting techniques an acrylic bulkhead with a hollow center can be used to secure the LED or detector in place. A recessed set-screw can be used to anchor the diode within the mount. In a typical application, a 14-mm, f/1 lens was mounted with epoxy. A 1-mm ring around the lens was required for the epoxy and the metal mounting tube. The result left the lens with an active diameter of 12 mm and a speed of f/1.2. The slight increase in f/number did not adversely affect the ability of the lens to collimate an infrared LED beam.

Filters, beam splitters, and other optical devices can be mounted using similar techniques. A major problem can occur, however, because of the widely varying sizes of commercial optical components. Small filters can sometimes be epoxied directly to the detector bulkhead so long as the cement is not allowed on the active area of the filter. Beam splitters can be epoxied to the ends of tubes cut at the desired angle (usually 45 degrees). For temporary applications use water-soluble casein cement so optical elements can be conveniently removed at a later date.

A good source of mounting tubing for lenses and diodes is hobby shops that market aluminum, brass, and plastic tubing. The tubing is economical and is available in closely fitting telescoping sizes. A rotary hand grinder can be used to cut the tubing. To avoid collapsing thin-walled tubing while cutting, use a wood dowel insert and clamp the tube in a padded vise.

Other lens-mounting techniques are available, and perhaps the best of all is to use threaded inserts. Threaded mounting is more difficult to accomplish than most other techniques. For details on these and other optical-mounting methods, consult appropriate references on optics.

Index

M

N

O

P